SWEET'NER DEAREST

BITTERSWEET VIGNETTES ABOUT ASPARTAME (NUTRASWEET®)*

H.J. ROBERTS, M.D.

CARTOONS BY

William Boserman
Pat Crowley
Bill Dwyer
Martin Filchock

Sidney Harris
Jeff MacNelly
Kevin McCormick
Nicholas Petkas, M.D.

Don Wright

* A registered trademark of The NutraSweet Company

i

Roberts, H.J.
Sweet'ner Dearest: Bittersweet Vignettes
About Aspartame (NutraSweet®)/by H.J. Roberts

Includes bibliographical references by author, and index.
ISBN 0 - 9633260-1-5 : $19.95
Library of Congress Copyright Number TXu 494 769

Library of Congress Catalog Card Number 92-64347

Printed in the United States of America

The Sunshine Sentinel Press, Inc.
P.O. Box 8697
West Palm Beach, FL 33407
(407)832-2409
FAX (407)832-2400

CONTENTS

DISCLAIMER

None of the observations, anecdotes, references or opinions cited in this book have been motivated by disrespect or malice toward the manufacturers, producers or representatives of products containing aspartame, saccharin and other products mentioned. W.E. Swirl, Inc. and WESINC are fictitious names.

QUALIFIERS

The term "aspartame products" used throughout the book refers to aspartame-sweetened products.

The term "aspartame disease" refers to illness associated with the use of products containing aspartame.

Bittersweet: "Bitter and sweet: pleasant yet painful"
Webster New Collegiate Dictionary

If anything can go wrong, it will.
Murphy's Law

Whatsoever is the father of disease, an ill diet was its mother.
George Herbert

You must still be bright and quiet
And content with simple diet......
But the unkind and the unruly,
And the sort who eat unduly,
They must never hope for glory -
Theirs is quite a different story!
Robert Louis Stevenson
(Good and Bad Children)

DEDICATION

This book is dedicated to these other
members of "Truth-Seekers Seven"

Barbara Mullarkey
Professor Janet Smith
Mary Stoddard
Dr. Paul Toft
Gailon Totheroh
James Turner, Esq.

PREFACE

The publication of my book, ASPARTAME (NUTRASWEET*):IS IT SAFE? (The Charles Press, Philadelphia, 1989), triggered national interest for several reasons.

- Over 100 million persons in the United States were consuming aspartame-containing products at the time.
- The Food and Drug Administration (FDA) was being inundated by an unprecedented number of complaints from consumers who attributed severe medical and other problems to them.
- The safety of aspartame products became a "hot" and controversial subject...as might be expected when the public's sweet tooth is threatened.

There lingered a nagging reservation about this "serious" book, however, after its printing. This stemmed from having omitted many humorous instructive vignettes encountered during "Project Aspartame." So consider the present opus not only its sequel, but also a balm to my conscience. Ogden Nash nicely put his finger on the matter in "I'm a Stranger Here Myself" (Inter-Office Memorandum):

"There is only one way to achieve happiness on this terrestrial ball,
And that is to have either a clear conscience, or none at all."*

Since this work largely consists of anecdotes, the reader might be interested in the term's origin. It derives from the Greek word "anekdotos," meaning "not given out." Procopus, a Byzantine historian in the 6th Century, used it as the title of a private collection of spicy gossip about public figures.

Other writers and editors have found the topics of sugar substitutes and presumed obesity to be fertile material for humor.

- The New Yorker (August 17, 1987, p.72), noting a statement in the Allentown (PA) Morning Call wherein researchers expressed the belief that aspartame can do more chronic damage than fed-

---

* From Verses From 1929 On. ©1932, 1935 Ogden Nash. Little, Brown and Company

eral health officials, offered a one-word commentary: "Nonsense." (My investigations
could support either position.)
- Reflecting on the current obsession with thinness and fad diets, satirists have suggested that
the sleek, as well as the meek, are likely to in-
herit the earth.

A story about Robert Benchley is appropriate. While attending
Harvard, this budding humorist was asked to discuss a fishing treaty.
Benchley replied, "I don't know anything about that treaty, but I'd like
to comment on it from the viewpoint of the fish." For your orientation
to this book, Dear Reader, just substitute "consumer" for "fish."

I consider the delivery of good humor by a practising physician
to be as challenging as writing medical texts for several reasons.
Others recognize the challenge.
- Sydney J. Harris correctly observed: "Humor is
as elusive a subject to pin down as love, or justice, or happiness" (The Miami Herald May 21,
1986, p. A-23).
- Christopher Fry noted in The Lady's Not for
Burning that laughter approaches the surest
touch of genius in creation.

First, humor tends to become an initial casualty when things go
berserk, whether in one's life or in the world. And yet, the paradox
remains that most encounters are truly humorous only within the context of real events and problems. In the present setting, the "bottom
line" of reality has the familiar ring of caveat emptor ("buyer beware.")

Second, this type of project could be risky business for a scientifically
oriented doctor such as myself. Perhaps "medical humor" falls in the
same category of oxymorons as "humble physician." (This may account
for William Osler's resort to the pseudonym Egerton Y. Davis in writing
a humorous case report.) Once again, Ogden Nash provided ample
warning in Don't Grin, or You'll Have to Bear It.

"It is better in the long run to possess an abscess
or a tumor
Than to possess a sense of humor.

People who have senses of humor have a very
   good time,
But they never accomplish anything of note, either despicable or sublime,
Because how can anybody accomplish anything
   immortal
When they realize they look pretty funny doing it
   and have to stop to chortle?"*

Third, I have misgivings about being called a "humorist" or "the
medical profession's Woody Allen"...descriptions that probably will
precipitate convulsions of laughter among my colleagues.   Even
Woody has personal reservation, as evidenced by his comment, "I think
that being funny is not anyone's first choice" (The Miami Herald April 8,
1992).

This effort at tickling your funnybone — or rib, if that's your ana-
tomical preference — could be regarded as "therapy" because laughter
is widely regarded as good medicine. Moreover, doctors must try to
maintain a sense of humor when assisting the physically and mentally
ill both for their sake and that of their patients.

- An Irish proverb states: "A good laugh and a
  good sleep are the best cures in the doctor's
  book."
- Aristotle (384-322 B.C.) commented: "The gods
  too are fond of a joke."
- Physicians, nurses and hospital administrators
  are being encouraged to utilize humor. Charles
  Schulz's Peanuts strip gave this advice to a character who wasn't feeling well: "Maybe you need
  more humor in your life" (The Miami Herald
  June 8, 1990).
- Many hospitals have "laughmobiles" or "comedy carts" stocked with funny VCRs, "humor
  rooms," and even a closed-circuit "humor channel." Some encourage visits by persons dressed

---

* From Verses From 1929 On. ©1932, 1935 Ogden Nash. First appeared in the New
Yorker. Reproduced with permission of Little, Brown and Company.

as clowns.
- The University Hospital in Albuquerque has its HAHAHA (Humor and Hospitals Are Healthy Allies) program.
- There's even an American Association for Therapeutic Humor (AATH).
- Dr. Jack Coulehan prescribes "the incredible lightness" as an "antidote for the heaviness of illness" (The Pharos Fall 1990, pp. 15-17). He fantasized on these enlightened ploys by hospitals: (1) a Laughter Clinic advertising, "A Laugh a Day Keeps the Scalpel Away"; and (2) sponsorship of a Comprehensive Poetry Program having the motto, "Sticks and Stones Can Break Your Bones, But Metaphor Can Heal You!"

Doctors should be encouraged to talk or write about their humorous encounters. My friend, Dr. Joseph D. Wassersug, underscored this effort in Medical World News (January 23, 1989). He bemoaned the dearth among his contemporary colleagues of a counterpart to the great humorist and satirist, Francois Rabelais (1495-1553) — also a physician — "...who, by his experience and insights, can enable the medical profession to enjoy a little laughter at its own expense."

Dr. Clifford Kuhn, a Louisville psychiatrist, seems to have carried "the frivolous pursuit" of medical humor to an ultimate. He performs as a stand-up comic on a circuit of clubs. Much of his material pokes fun at doctors, including himself. For instance, one involves telling his receptionist that he is too busy to see a man in the waiting room claiming to be invisible (The Wall Street Journal September 10, 1991, p. B-1).

Advocates of "humor therapy" understandably go to lengths to indicate its physiological basis. To illustrate, a good belly laugh can stimulate endorphins (natural mood elevators and painkillers) or prevent their depletion by the hormones that stress provokes. It also provides aerobic exercise by increasing the pulse, muscular activity and breathing, and even may aid the immune system.

*"Can you explain it to me in something
simpler than 'layman's terms'?"*

Cartoon P-1
© 1963 Medical Economics Company. *Reproduced with permission.*

But the serious punchline of this volume constitutes its essential justification — namely, educating consumers about a complex medical and public problem in a way that bypasses "medicalese." Cartoon P-1 illustrates this theme.

Admittedly, I have included a few medical perspectives because of their relevance. Examples are aspartame reactions afflicting multiple family members, and the pervasive "fear of fat" that is conducive to severe eating disorders.

Another occupation that demands the preservation of one's sense of humor is the cartoonist. With a few strokes of the pen and a paucity of words, the talented cartoonist can define a universal reality or sentiment that is promptly recognized by readers. Not surprisingly, I sought out the wonderfully honed skills of several cartoonists for enhancing the enjoyment and educational impact of this book.

So enjoy...and please pay attention to the messages accompanying the chuckles. They could be far more cost-effective than the psychiatric and other medical encounters of patients who are described in these vignettes.

H. J. Roberts, M.D.
West Palm Beach

ACKNOWLEDGEMENTS

I am grateful to Shirley Brightwell and Kathleen Brightwell for their secretarial services.

Permission was graciously given for the reproduction of "Who Killed The Sugar Plum Fairy?" by Mary Nash Stoddard, "To Hell and Back on Aspartame" by Kelly Allen, and "Case 552" by Mary Lou Williams.

Beatrice Trum Hunter, Esther Sokol and Stephen Roberts provided many valuable suggestions for the manuscript.

I am indebted to Helen Musgrave, my office nurse for more than 30 years (!), whose good humor preserved mine on many an occasion.

The following persons and companies granted permission to reproduce excerpts:

American Medical Association
Dow Jones & Company, Inc. (The Wall Street Journal)
The Journal of the Florida Medical Association
The Miami Herald
The Palm Beach Post
Professor Darden Asbury Pyron
Dorothy Schultz (President, Hypoglycemia Association, Inc.)
Harry Sewall, JD (Editor, Expert Witness Journal)
Southern Medical Journal and John B. Thomison, M.D. (Editor)
Charles C Thomas Company (publisher of Diseases of Medical Progress: A Study of Iatrogenic Disease by Dr. Robert H. Moser)
Ann Topper
Wednesday Journal

I was fortunate to enlist the talents of my colleague, Nicholas Petkas, M.D., whose brilliant cartoons have delighted me for many years.

These artists, syndicates, organizations and companies gave permission to use cartoons.

American Diabetes Association (cartoon by
Mr. Damon Hertig (in Diabetes Spectrum)
Mr. William H. Boserman

Mr. Pat Crowley
Mr. Bill Dwyer
Mr. Martin Filchock
Mr. Sidney Harris
Mr. Kevin McCormick (creator of <u>Arnold</u>)
Medical Economics Book Division, Inc. (publisher of <u>Cartoon Classics</u>)
The Medical Tribune Group
News America Syndicate
Tribune Media Services (<u>Shoe</u> cartoons)
Mr. Don Wright

I also am grateful to Little, Brown and Company for permission to include a number of excerpts from poems by Ogden Nash that enhance both the meaning and humor of this book.

AN OVERVIEW OF ASPARTAME AND REACTIONS TO ASPARTAME PRODUCTS

I really intend to keep my promise about minimizing medicalese. Since this is a health-related subject, however, I owe the interested reader an introductory overview of both aspartame and reactions to products containing it. More details can be found in the listing of my publications on this subject (pp. 291-293).

ASPARTAME

The Food and Drug Administration (FDA) approved aspartame, a low-caloric sweetener, for use in solid foods during 1981, and in soft drinks during 1983.

Aspartame is a synthetic chemical. Stated differently, it is neither "natural" nor "organic." Aspartame consists of phenylalanine and aspartic acid — two amino acids ("the building blocks of protein") — and a methyl ester. The latter promptly becomes transformed in the stomach to methyl alcohol, also known as methanol or wood alcohol...an unequivocal poison. It represents 10.9 percent (!) of the aspartame molecule.

Senior FDA scientists vigorously protested the licensing of aspartame-containing products for nearly a decade prior to its approval. They and others based their objections on many disturbing findings in animal studies (especially brain tumors), seemingly flawed experimental data, and the absence of extensive pre-marketing trials on humans.

> Congress approved a food additive amendment in 1958 intended to protect the public against unsafe food additives to the same degree as unsafe drugs. Its objective was stated in Congressional and Administrative News, Legislative History of Food Additives Amendment of 1958 (p. 5302): "Safety required proof of reasonable certainty that no harm will result from the proposed use of an additive." It amplified the "concept of safety" for food additives in these terms:

"The concept of safety used in this legislation involves the question of whether a substance is hazardous to the health of man or animal. Safety requires proof of a reasonable certainty that no harm will result from the proposed use of an additive."

ASPARTAME CONSUMPTION

Aspartame captured the public's tastebuds. Its sweetness is estimated as 180 times that of table sugar. A sophisticated PR barrage, coupled with negative consumer attitudes toward saccharin as the result of mandated labeling about possible cancer (based on findings in a few rats, but still not confirmed in animals or humans given reasonable amounts) propelled its popularity.

Well over 100 million persons in the United States currently consume about 4,200 products containing aspartame (Chapter 20). It is widely known as NutraSweet® and Equal®, registered trademarks of The NutraSweet Company.

This "marketing miracle of the 1980s" continues. For example, the FDA approved aspartame incorporation in juice drinks, frozen novelties, tea beverages, breath mints, yogurt-like products, refrigerated flavored milk beverages, ready-to-serve fruit juices, refrigerated ready-to-serve gelatin desserts, fruit wine beverages and frozen desserts (see 20). The commercial spiral promises to accelerate. Approval has been obtained for the addition of aspartame to baked goods (using an encapsulated form), low-alcohol beer, and a host of other products. Furthermore, a taste-alike and "guilt-free" artificial ice cream, consisting of aspartame and a low-calorie "fake fat," is on the market (see 43).

The economic aspects of this phenomenon are unprecedented. A previous chairman of The NutraSweet Company estimated that the foregoing new markets could generate total annual sales of $6 billion!

PROFESSIONAL CONCERN

As a physician who treats many patients with diabetes, hypoglycemia ("low blood sugar attacks"), and related metabolic medical problems, I initially rejoiced over the availability of aspartame-containing products. After all, here was a palatable sweetener devoid of calories, sugar, salt and cholesterol.

This view changed radically following first-hand clinical observations and nationwide correspondence. As of January 1992, I had data on 632 persons with serious — and occasionally life-threatening — reactions to aspartame products. The majority completed my detailed 9-page survey questionnaire. (Copies are available without charge to any interested reader.)

There have been a few independent reports by other physicians. Most medical journals, however, are reluctant to publish such observations on the grounds they are "anecdotal" reports of "idiosyncratic" reactions, or not sufficiently "scientific."

The counterproductive nature of such medical literary elitism and regulatory arrogance has surfaced in several ways. Perhaps the most impressive is the fact that the FDA received complaints from more than 5,900 consumers as of July 1, 1991. They represented 80 percent (!) of the reported adverse reactions to food ingredients since this agency's monitoring program began in 1985 (Chapter 3). They included 473 incidents of epileptic convulsions and other seizure-like disorders!

Many other patients and consumers have registered legitimate complaints to physicians, manufacturers, and consumer groups (e.g., Aspartame Victims and Their Friends; the Community Nutrition Institute). But most doctors simply throw up their hands after being told that aspartame represents "the most tested and safest additive in history."

REACTIONS TO ASPARTAME PRODUCTS

My criteria for the diagnosis of an "aspartame reactor" are straightforward. They include the following: (1) the disappearance of complaints, or gratifying relief, within several days or weeks after avoidance of such products; and (2) their prompt and predictable recurrence on resuming aspartame consumption. The latter involved repeated trials (over 10 times by many individuals) and provocative testing. For example, I induced a seizure in a 16-year-old female having recurrent and previously-unexplained epilepsy within three hours after she swallowed just one serving of an aspartame chocolate pudding in my laboratory. It was given (with informed consent) following a seizure-free period when aspartame products were avoided.

The <u>most frequent and serious reactions</u> in my series are summarized as follows:

- <u>Severe headaches</u> — nearly half (specifically, 47 percent)
- <u>Severe dizziness and instability</u> — over one-third
- <u>Marked irritability, severe anxiety attacks, gross personality changes, the aggravation of phobias</u>, or combinations thereof — over one-third
- <u>Decreased vision or actual blindness</u> (one or both eyes) — over one-fourth
- <u>Profound confusion and memory loss</u> — one-fourth
- <u>Extreme depression</u> (often with suicidal thoughts) — one-fourth
- <u>Epileptic attacks</u> (grand mal, petit mal, psychomotor attacks) — one-fifth

<u>Other frequent complaints</u> read like a medical textbook. (This argument actually has been used by my corporate critics.) They include <u>itching, hives, other skin rashes, irritation or swelling of the lips, mouth or tongue, debilitating fatigue, intense sleepiness, hyperactivity, uncontrollable tremors, severe insomnia, menstrual changes in young women</u> (ranging from the loss of periods to frequent and heavy bleeding), <u>difficulty in swallowing, excruciating abdominal pain, diarrhea</u> (at times bloody), <u>intense thirst, "dry eyes"</u> (often aggravated by wearing contact lens), <u>ringing of the ears, decreased hearing, marked sensitivity to noise, palpitations, recent hypertension, atypical chest pain, frequency or burning on urinating, the aggravation of respiratory and other allergies, the precipitation of clinical diabetes, loss of diabetic control, unexplained increase in the severity and frequency of hypoglycemic attacks, joint pains, recurrent infections, a paradoxic *gain* in weight</u> (as much as 80 pounds), <u>severe weight loss</u> (see below), <u>and drug interactions.</u>

Furthermore, most of my patients had <u>multiple</u> (5 or more) reactions to aspartame products. This was evidenced by the prompt subsidence of such symptoms after abstinence.

Most doctors are not aware of these associations. Others choose not to believe their legitimacy for reasons already noted. I must further temper criticism of my colleagues for additional considerations. First, the safety of aspartame products continues to be enthusiastically reaffirmed by the FDA. Second, the manufacturer will promptly inundate any physician inquiring about reactions with scores of "scientific" articles and hundreds of references seemingly supporting such safety. But there's a hitch: few practitioners have the intense interest or academic credentials to challenge the data in such reports, the vast majority of which were funded by corporate interests.

There is an obvious corollary. <u>Every</u> consumer of aspartame products with medical, psychiatric, behavioral or other problems that remain unexplained should read this book — or recommend it to afflicted relatives and friends who consume such products. A few excerpts from the considerable correspondence I receive illustrate the value of such advice.

- "I couldn't drive for three months. I couldn't travel for fear of a seizure. I delayed the start of a business, and couldn't complete old business. Anxiety — never knowing when a seizure might occur, or why medication wasn't working." (31-year-old educational consultant with seizures following use of aspartame products)
- "If it wasn't for the Consumer's Report Special, I might have never associated my severe headaches with the aspartame. Only God knows what would be wrong with me now. The night I saw the show, I immediately stopped drinking diet drinks. The next night the headaches were gone." (24-year-old woman)
- "Before all this happened, I was a very active 48-year old woman. I played golf almost every day and at least twice a week played 36 holes a day. I also did

my own yard work and my neighbor's, which included eight fruit trees, two palm trees and countless shrubs. I went from that to no yard work and only occasionally tried to play golf...I want to stress the fact that this was a daily battle and not just occasionally. I went from one symptom to many, and the mental anguish and depression were devastating. I do mean it when I say I felt like I was always trying to recover from a hangover. The bottom line: since I quit using aspartame, I am back to playing golf, doing yard work, and feeling wonderful."

HIGH-RISK GROUPS

The problem of reactions to aspartame products assumes even greater importance for certain high-risk groups. The most notable are children, pregnant women, and patients with migraine, diabetes, hypoglycemia, and epilepsy.

Additional groups include nursing mothers, relatives of aspartame reactors (see below), older persons (especially when memory loss already exists), allergic individuals, patients with hypothyroidism (underactive thyroid), and persons prone to phenylketonuria (PKU).

One out of 50 persons in the general population harbors the PKU trait due to deficiency of an enzyme required to metabolize phenylalanine. The ensuing accumulation of phenylalanine from food in a pregnant PKU carrier can cause severe mental retardation and other problems during fetal life and infancy.

There are other categories of vulnerable consumers. They include persons with alcoholism, iron-deficiency anemia, poor kidney or liver function, and patients taking drugs known to interact with phenylalanine. Examples of the latter are L-dopa for Parkinsonism, and methyldopa for hypertension.

FAMILY INCIDENCE

An unexpected, but consistent, finding has been the frequent occurrence of reactions to aspartame products in close relatives. Indeed, one out of five aspartame reactors could identify two to seven family members who were afflicted with this problem. It is of added interest that they often exhibited the same manifestations.

THE TOLL

The costs for diagnostic studies, consultations and hospitalization incurred by patients with unrecognized reactions to aspartame products have been high; in some instances, they were astronomical. For example, the family of a child wrongly suspected of having leukemia (on three separate occasions) estimated such expense at $750,000 (!) before the role of aspartame products was suspected.

Many patients were subjected to invasive testing such as carotid or coronary angiography. Some underwent needless surgery. This conclusion is based on the persistence of complaints postoperatively, and their subsequent prompt regression after stopping aspartame products. The operations included removal of cataracts for visual complaints, prostate surgery (transurethral resection) for urinary symptoms, and curettage of the uterus for excess bleeding.

The toll also was profound in terms of occupational loss, economic devastation, and personal incapacitation. For example, a number of "reactors" had their driver or pilot licenses revoked, notwithstanding the subsequent absence of seizures and other complaints after avoiding aspartame products.

MECHANISMS

Damage to the retina, optic nerves and retina can be caused by methyl alcohol. As noted above, this poison is promptly released after ingesting aspartame.

Other complications are attributable to the high levels of phenylalanine and aspartic acid within the brain. They can alter amino acid-derived neurotransmitters (chemicals that regulate nerve and endocrine function), and affect the release or metabolism of insulin, growth hormone and other important substances. Many of the clinical features mentioned earlier reflect the brain's limited ability to cope with recurrent massive flooding by these individual amino acids.

17

Concomitant <u>severe restriction of calories</u> can have equally disastrous effects on the brain, peripheral nerves, heart and other organs. This is especially worrisome for figure-conscious women afflicted with "fear of fat" or "five-pound-overweight paranoia" (Chapter 24). Some young women in my series were diagnosed as having anorexia nervosa and bulimia after losing up to 90 pounds.

PUBLIC HEALTH IMPLICATIONS

More than 100 million persons in the United States currently consume products containing aspartame. Even if only <u>one-tenth of one percent</u> reacted adversely to this chemical, an exceedingly low figure in my opinion, over <u>125,000</u> would still be at risk!

When dealing with reasonable inferences of this magnitude, the "anecdotal" assertions by those who express such concerns are valid, and not exaggerated imagination. After all, how could a single physician have compiled data on over 600 persons meeting the foregoing criteria of "aspartame reactors" if an enormous submerged iceberg of comparable problems did not exist in the population? The additional offensive implication that these perceptive individuals are "just a bunch of hypochondriacs" could be regarded as a self-serving tactic.

Assume that my projections might be slightly overstated in a worst-case scenario. Dare we wait for undeniable confirmation of an epidemic involving seizures, mental retardation, dementia, behavioral problems, suicide, birth defects, and a host of new or aggravated medical problems? Addressing the possibility of such an imminent public health hazard before the Section on Internal Medicine of the Southern Medical Association on November 10, 1986, I urged a prompt in-depth reassessment by corporate-neutral investigators using acceptable research protocols. This plea fell on deaf ears.

The FDA's <u>imminent hazard regulation</u> (21 CFR 2.5) is pertinent.

> "(a) Within the meaning of the Federal Food, Drug
> and Cosmetic Act, an imminent hazard to the pub-
> lic health is considered to exist when the evidence
> is sufficient to show that a product or practice,

posing a significant threat of danger to health, creates a public health situation (1) that should be corrected immediately to prevent injury and (2) that should not be permitted to continue while a hearing or other formal proceeding is being held. The 'imminent hazard' may be declared at any point in the chain of events which may ultimately result in harm to the public health. The occurrence of the final anticipated injury is not essential to establish that an 'imminent hazard' of such occurrence exists.

"(b) In exercising his judgment on whether an 'imminent hazard' exists, the Commissioner will consider the number of injuries anticipated and the nature, severity, and duration of the anticipated injury."

As cases in point, the potential for tumors (Chapter 34) and for birth defects being induced by aspartame products requires clarification. Two Senate hearings — held on May 7, 1985 and August 1, 1985 — detailed the concern about tumors by FDA scientists. Since the 5-year statute of limitations for challenging two pivotal tests had been allowed to expire as a result of seemingly intentional footdragging, the Delaney Amendment to the Food, Drug and Cosmetic Act could not be invoked. Dr. M. Adrian Gross, a senior FDA pathologist-scientist, testified before the Senate hearing on August 1, 1985:

"In view of all these indications that the concern-causing potential of aspartame is a matter that has been established way beyond any reasonable doubt, one can ask: 'What is the reason for the apparent refusal by the FDA to invoke for this food additive the Delaney Amendment to the Food, Drug, and Cosmetic Act?'"

My studies raise additional disturbing questions. Here are a few.

- Can selective amino acid imbalances of phenylalanine and aspartic acid, resulting from the prolonged ingestion of aspartame products, accelerate or aggravate Alzheimer's disease, Parkinson's disease, epilepsy, migraine, altered behavior, intellectual deterioration, and other neuropsychiatric or metabolic afflictions?
- Should labeling be required to warn drivers, pilots, and persons with eating disorders (obesity; anorexia nervosa; bulimia) about the risks associated with use of aspartame products (Chapter 24)?
- Are the earlier animal studies indicating a high incidence of brain tumors in aspartame-fed rats that were ignored by an FDA commissioner now coming home to roost in view of the dramatic rise of human brain cancer following the availability of aspartame products (Chapter 34)?

ANOTHER PANDORA'S BOX

Another related issue deserves considerable emphasis. Aspartame heads an expected avalanche of synthetic foods, substitute foods and additives that are being created by food technologists through ingenious physical and chemical manipulations. Their long-term biological consequences might prove devastating, however, if released as "generally recognized as safe" (GRAS) without the results of extensive premarketing studies on humans...as occurred with aspartame!

My sentiments coincide with those of Senator Howard Metzenbaum. He stated in the Congressional Record - Senate of August 1, 1985:

> "We had better be sure that the questions which have been raised about the safety of this product are answered. I must say at the outset, this product was approved by the FDA in circumstances which can only be described as troubling."

A PHYSICIAN'S PLEA

I could not have predicted that four decades of medical practice and clinical research would culminate in being designated as a "majority of one" within this context. The reason for electing to become a consumer advocate therein is nicely embodied in a phrase by Stanley S. Arkin: "...standing up to authority that is unwisely, arrogantly or selfishly exercised serves a higher good"* (The Wall Street Journal March 13, 1990, p. A-16).

Scores of aspartame "victims" and correspondents encouraged me to continue this effort in unsolicited letters.

- A Florida registered nurse wrote: "I admire your courage and integrity in tackling aspartame. Thanks for being a dragon-slayer."
- A New York correspondent opined: "This is to thank you for your recent book on aspartame. Your profession owes you theirs, too, for standing up for honest and conscientious medical ethics in denouncing this travesty on the health and health costs of our and other nations."
- A woman from California noted: "It is wonderful that you have the courage and perseverance to continue your research in the face of all the opposition from those with vested interests in products containing aspartame."
- A Louisiana resident stated: "I want to thank you for speaking out about aspartame causing medical problems...When a mere patient tries to speak out to defend his health against chemical additives, he meets deaf ears and closed doors. But when a doctor such as yourself is brave enough to speak out for us, we are encouraged that there may yet be a chance to get the chemicals out of our food. Thanks again. Please don't let 'them' silence you."

The validity of my observations and inferences is underscored by the fact that they were corporate-neutral. This means I received no grants from companies or organizations having any vested interest in this area.

It is necessary to sensitize health-care professionals and consumer groups about (1) the potential adverse effects of aspartame and related products, and (2) the harm that could result from reflexive bureaucratic and corporate footdragging in such matters. Lest we forget, it took more than five years to prove that thalidomide really was the cause of severe birth defects — and even longer to link cigarette smoking with lung cancer.

Congress and the FDA must address the following serious deficiencies concerning the evaluation, licensing, promotion and surveillance of aspartame-containing products to avert a potential "recipe for disaster."

- The paucity of accurate clinical and epidemiologic data in aspartame consumers, especially brain and urinary bladder tumors. Dr. Douglas L. Park, Staff Science Advisor for the Office of Health Affairs of the Department of Health & Human Services, concluded his 1981 presentation of the Public Board of Inquiry's hearing on aspartame safety relative to brain gliomas in these terms:

  "I believe that aspartame has not been shown to be safe for the proposed food additive uses. Along with the Board of Inquiry, I must recommend, therefore, that aspartame not be approved until additional studies are carried out using proper experimental designs."

- The failure to challenge the alleged safety and effectiveness of aspartame in current mass promotional campaigns, especially for limiting carbohydrate intake and long-term weight loss. Many dieters and diabetics actually increase their intake of sugar and fat while using aspartame products.

- The acceptance of "negative" double-blind studies funded by corporate interests that fail to reproduce reality. Special attention is directed to the administration of aspartame as capsules or in freshly-prepared cool solutions rather than the implicated commercial products. Such seemingly "scientific" investigations nicely sidestep the issues of multiple stereoisomers (mirror images) of aspartame's amino acids, and other breakdown products formed during heating or prolonged storage. When an independant double-blind controlled study of epileptic children used an aspartame product "obtained from a grocery store," the frequency of their spike-wave discharges (by EEG) increased (Neurology Vol. 42:1000-1003, 1992).

- The severe shortcomings of tests on experimental animals that are inherent in their vast differences from man relative to metabolism and life span. For example, the apparent failure of rats to perceive aspartame as "sweet" might not trigger the clinically important cephalic (brain) phase of insulin release.

- The FDA's maximum daily allowable intake (MAI) of 50 mg aspartame/kg, a figure I regard as arbitrarily excessive.

- The resistance by Congress to mandate proper labeling in terms of dating these products, specifying the quantity of aspartame they contain, and warnings about prolonged storage and exposure to heat.

- The unfounded ongoing bias against saccharin, based on limited and highly controversial studies involving urinary bladder tumors in a few male rats. ("Use of this product may be hazardous to your health. This product contains saccharin which has been determined to cause cancer in laboratory animals.") Furthermore, testing for such tumors after aspartame administration in several key studies was reported only in female rats, which are more resistant than males to develop them.

- The failure of the FDA to issue warnings based on thousands of major medical, neuropsychiatric and reports of behavioral side effects received from consumers.

WILL HISTORY REPEAT ITSELF?

The serious public health, scientific and political ramifications of this "exposé" could be unprecedented. The FDA's continued approval of aspartame products in the face of thousands of <u>volunteered</u> complaints from consumers — including hundreds of cases of epileptic disorders — warrants expressions of outrage.

Consumers and consumer advocates who have involved themselves in the problem of reactions to aspartame products can attest to the magnitude of corporate denial, professional indifference, and publisher intimidation. Given the huge PR budgets used to promote aspartame products, one can understand the kill-the-messenger attitude held by most advertising executives of both professional and popular magazines. It is encapsuled in the command: "Kill that story!"

A historical twist is germane to the orientation of this book. After a frustrating ordeal, one school teacher who had reacted to tea containing aspartame suggested, "What we need is another Boston Tea Party!"

1

PATIENT ENCOUNTERS

Many patients and correspondents provided one-of-a-kind "funny" stories about their experiences with aspartame products.

The Wrong-Person Hangover

I had treated a woman for hypertension over many years. She and her husband visited the neighborhood social club on weekends. As he enjoyed his whiskey, she sipped an aspartame-containing soft drink. This patient complained: "The problem is that I'm the one who gets the hangover...not him!" Her "hangovers" ceased as soon as she avoided "diet" beverages.

Cartoon 1-1

Aspartame Lung?

An aspartame reactor improved when she discontinued using such products. The woman later returned because of a cough. She jokingly self-diagnosed the condition as "aspartame lung" because her husband continued to take large amounts A.W.A. — against wife's advice.

Increasing Business?

Several aspartame reactors who were hospitalized for other reasons vehemently objected to being served aspartame products. All asked virtually the same question: "Doctor, why in the world do these hospitals continue to serve aspartame? Aren't they supposed to be places to go for regaining one's health?"

One patient had a cynical explanation: "Of course, it could be a way of increasing business."

Sound-Alikes

A spirited patient had avoided aspartame products...with considerable relief of several complaints. This elderly lady also was slightly deaf.

I concluded a subsequent visit by saying, "You're a delight." She replied, "But I don't drink those 'lite' beverages any more!"

Information From the Track

It's been a while since I attended a horse race or even read about one. If asked to name any recent triple crown winner in a game of <u>Trivial Pursuit</u>®, I'd draw a blank.

It was quite a surprise, therefore, to find the October 1986 issue of <u>Post Time USA</u> on my desk. This 13th Anniversary Edition contained 80 pages of horse racing news. Attached to the cover was an intriguing note from a patient: "Dr. Roberts, please see page 5 and page 49. You turn up in the most unexpected places!"

Flipping to page 5, I found a feature about my researches on aspartame. And as a bonus, page 49 contained the <u>entire</u> release of my recent press conference. Moreover, it was printed against a gray background and surrounded by a heavy border for editorial emphasis. Wonderment...

For me, though, the humorous punchline appeared on the <u>opposite</u>

page. It concerned a physician's recommendation given Henny Youngman about cutting his sexual activity in half. The comedian then asked the doctor which half he should give up: "...talking about it, or thinking about it?"

2

OTHER ENCOUNTERS

The Unsuspecting Secretary

Elaine is a fabulous secretary. Having typed numerous case reports and scientific entries for "Project Aspartame," she was quite familiar with the subject. Not surprisingly, she studiously attempted to avoid such products.

Elaine came into my office one morning wearing a stern expression. Clutching a box of "extra fiber" cereal, she exclaimed, "Doctor, did you know that this contains aspartame?" The oversight apparently had occurred because large promotional printing on the container dominated the list of ingredients. Furthermore, the sweetener's emblem was not displayed. Elaine then confessed, "It's hard to believe that I would be taking aspartame in any of my food!"

Such concern over the labeling, or mislabeling, of food products offers cartoonists plenty of fodder. Considering the frequent long lists of ingredients, one placed this label: "Ingredients: Do you really want to know?"

Cartoon 2-1

Visiting Relatives and The Unknown Boss

Betty had been my insurance clerk for many years. Several days after the foregoing revelation by Elaine, Betty announced that relatives from Pennsylvania were visiting.

This didn't seem an earthshaking announcement. Betty then mentioned the _first_ question one had asked: "Do you happen to know a Dr. Roberts in this area? We're very interested in the research he's doing on aspartame." She thereupon stated, "Well, it just so happens that I work for _that_ Dr. Roberts!"

A Spouse's Encounter

A local TV station featured my first press conference (see Chapter 23) on the 6:00 P.M. news.

One week later, my wife went to the hairdresser. Carol noticed that he was sipping an aspartame-free soda. As soon as she mentioned the subject, this chap — a fellow in his 30s — exclaimed, "Is _that_ Doctor Roberts _your_ husband?" When she replied in the affirmative, he confided:

> "Let me tell you what happened to _me_! I was having lots of trouble, especially blurred vision when I watched television. It even occurred while I was working. When I heard him mention that these products could cause blurring, I stopped them immediately! My sight is much better now...honest."

Carol was pleased on two accounts: her hairdresser's improvement, and the fortunate delay of this appointment with him for more than one week due to an intervening urgent conference. She reflected that her husband's research may have prevented a scalping... literally.

Cartoon 2-2

More on Reading the Fine Print

Carol and I were dining at a local restaurant in our South Florida community. The animated conversation at an adjacent table was clearly audible. Two older couples seemed intensely interested in the detailed labeling of "the blue" (aspartame) and "the pink" (saccharin) packets of tabletop sweeteners.

One of the men then began reading the contents out loud. I marveled at the apparent scientific bent of these seniors.

But alas, the ensuing scenario once again confirmed that perception may not represent reality (see 10). The fellow doing the reading actually was trying to prove a point: he still could read fine print without wearing glasses!

Cartoon 2-3

Misperception of Pornography

Barbara Mullarkey (see 23), a nutritional columnist and publisher of NutriVoice, suggested that I see The Stuff. Recognizing her to be a straight-shooter who wouldn't recommend a book or video without good reason, I rented a copy and placed it temporarily on the dining table. Returning to the room 10 minutes later, I was accosted by strange gleams in the eyes of my wife and sister-in-law.

Wife:	"You got a porno film?"
H.J.R.:	"What makes you say that?"
Wife:	"Just look at the label of this video you rented!"

H.J.R.:	"What are you talking about?"
Sister-in-law:	"It says, 'Are you eating it, or is it eating you?'"
H.J.R.:	"So?"
Wife:	"And it's rated R!"
H.J.R.:	"I want you to know that Barbara Mullarkey recommended it. She's not only a columnist I respect, but also the mother of five children."
Sister-in-law:	"But that doesn't mean it's not pornographic."

Having gone through this third degree, I was even more interested in viewing The Stuff, a film written and directed by Larry Cohen. Not being a fan of "horror movies," I opted against partaking of popcorn.

The basic plot dealt with a remarkable better-than-ice-cream substance. In fact, it was so enticing that "enough is never enough." Weight-conscious consumers also were delighted by the minimal number of calories. Competitors who manufactured ice cream retained a private investigator to probe both its origin and approval by the FDA.

I won't give away the rest of the story other than to indicate that poetic justice occurred in the last scene when bigwigs selling this "natural" product were forced to eat "The Stuff"...at gun point. One, a previous competitor, seems to have followed the scenario of several soft-drink manufacturers who originally opposed the incorporation of aspartame into soft drinks, but then made that famous decision: "If you can't beat them, join them!"

An Encounter by the Office Staff

One of the office telephone lines became silent. My nurse called the phone company. A 33-year-old fellow arrived and repaired the problem. As he was about to leave, the announcement of my participation in a recent seminar on aspartame safety caught his eye.

Repairman:	"Does your doctor really think aspartame isn't completely safe?"
Nurse:	"You bet! He even wrote a book about it."

Repairman:	"What do you know!"
Nurse:	"Why do you ask?"
Repairman:	"Well, I break out into hives every time I use the stuff."
Nurse:	"Oh?"
Repairman:	"Yeah. I even called the company."
Nurse:	"What did they tell you?"
Repairman:	"They said it must be my coffee."
Nurse:	"So what happened?"
Repairman:	"The hives came out when I tried it in another brand."
Nurse:	"Want to do something about it?"
Repairman:	"Sure, if I can."
Nurse:	"Fill out Dr. Roberts' questionnaire... NOW!"

This dialogue promptly triggered Kelly Allen's reference to "my coffee" (Chapter 45).

3

CARTOON FUN

Astute cartoonists have found they can evoke chuckles by satirizing sugar-free products.

- The cartoon strip <u>Born Loser</u> for January 21, 1992 (<u>The Palm Beach Post</u>) focused on a husband being served coffee by his wife. She complained that he developed "artificial diabetes" after switching to a "sugar substitute" due to worry about the adverse health effects of too much sugar. (In my experience, the qualifier "artificial" could be omitted in the case of aspartame products.)
- Kevin McCormick hit the jackpot in his <u>Arnold</u> strip of October 2, 1986.

Cartoon 3-1

But the Roberts' Gold Star for the most panoramic cartoon yet seen goes to Bill Dwyer for his <u>Foibles</u> rendition in the Winter 1988 issue of <u>Conscious Choice</u>.

Cartoon 3-2

The FDA

The <u>FDA Consumer</u> is one place I hardly expected to find a frivolous cartoon on this subject. Yet the following one appeared in its October 1988 issue alongside this caption:

Adverse Reactions to Food Ingredients

FDA has received approximately 6,000 complaints about adverse reactions to food ingredients since it began its monitoring program in 1985. The artificial sweetener aspartame accounts by far for the largest number of complaints.

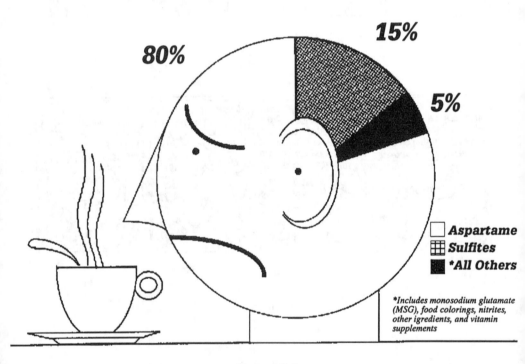

80%   **15%**   **5%**

☐ **Aspartame**
⊞ **Sulfites**
■ ***All Others**

*Includes monosodium glutamate (MSG), food colorings, nitrites, other igredients, and vitamin supplements

Cartoon 3-3

Solving Clinical Riddles

Several cartoon renditions could have served as illustrations for diagnostic problems caused by unrecognized reactions to aspartame products (see 33). As the author of DIFFICULT DIAGNOSIS: A GUIDE TO THE INTERPRETATION OF OBSCURE ILLNESS (W. B. Saunders Company, 1958), I can attest to such similarities.

My repeated encounter of "aspartame disease" fostered such appreciation of cartoons beyond the intent of their creators. Charles Schulz's classic feature in Peanuts, "Psychiatric Help 5¢: The Doctor Is In," provided several examples.

His strip in The Miami Herald for April 8, 1990 showed a "patient" with recurrent depression, and recalled the advice of a "psychiatrist" about always remembering that "every cloud has a silver lining." I fantasized: "If he happens to be a reactor to aspartame products, the 'silver lining' is obvious: avoid them. At the very least, it couldn't hurt."

Confusion and Memory Loss

Cartoons in newspapers, magazines, and the myriad of "throw away" periodicals sent to doctors often hint at the enormity of confusion and memory loss. For example, an executive seated behind a desk asked his secretary, "Have I eaten lunch yet?" This inference is even more amusing among health-conscious businessmen and professionals who are likely guzzlers of diet sodas.

But cartoons of this genre really hit a responsive note when aimed at doctors. Perhaps I was more attuned to the matter because dozens had flatly denied that confusion and memory loss, as well as other disorders, could be caused by aspartame products. One such

"Ms. Ames, what's the name of the lady with the memory loss problem?"

Cartoon 3-3
©1987 Medical Tribune. Reproduced with permission.

rendering in <u>Medical Tribune</u> showed a puzzled physician asking his nurse for "the name of the lady with the memory loss problem."

The Pop Culture Monster

Readers with diverse backgrounds are likely to react differently to the same cartoon. For example, Pat Crowley's offering in the March 13, 1987 edition of <u>The Palm Beach Post</u> depicted a monster gripping a girl as her parents chatted. (The word "monster," by the way, originates from the Latin "monere" or "monstrum" — meaning to warn, or a divine omen indicating misfortune.) The dragon's mouth included the term "Pop Culture." The father nonchalantly observed, "Our daughter just hasn't been herself lately." To which the mother replied, "What's

Cartoon 3-5

©*1987 <u>The Palm Beach Post</u>. Reproduced by permission of Mr. Pat Crowley.*

eating her?"

My initial reaction was almost reflexive: "Why not ask what she's eating and drinking?" This was the converse of a remark I had jokingly made over the years to many patients with peptic ulcer: "It's either something you 'et, or someone you met."

4

MORE ON "LITE" AND "NO-NOTHING" CARTOONS

The "charge of the lite brigade" is a contemporary phenomenon. Secretary of Health and Human Services Louis W. Sullivan expressed concern over the public's dazzlement by "light" foods. He envisioned grocery stores becoming a "Tower of Babel" in which consumers would have to become "linguists, scientists and mindreaders to understand many labels they see."

All the while, cartoonists have been reveling with "lite" foods and "no-calorie" beverages. Such scenarios reminded me of the famous line in Shakespeare's As You Like It: "Sans teeth, sans eyes, sans taste, sans everything." This "no-nothing" theme seemed particularly timely for business periodicals during the recent savings-and-loan "no-money" fiasco.

- Two laboratory scientists reflected on what was left in their drinks after "all the sugar, caffeine, fat and carcinogens" had been removed.
- A fellow at a bar wondered to his Martini-sipping companion why he even bothered drinking something that had no alcohol, no sugar, and no caffeine.

Rationalizing The Widespread Consumption of Diet Sodas

Bill Dwyer encompassed much of the concern about aspartame-containing beverages in his cartoon for the Winter 1988 edition of Conscious Choice. His important punchline in Cartoon 3-2 speaks for itself.

The comic strip Cathy provided another insight. The subject concerned rationalization — as when her group indulged in a feast of "doughnut holes" (The Palm Beach Post July 17, 1990). They enjoyed such excess because "someone has already rationalized it for us."

Cohen's Dictum

Some dubbed the success of aspartame as "the marketing miracle of the 1980s." But one aspect bugged economists. It was encapsulated in the dictum expressed by Stephen M. Cohen in The Journal of Irreproducible Results (Volume 35, No.1,1990, pp. 15-16): "The more you pay, the less you get" For those demanding more scientific lingo, Cohen's article was titled, "On the Inverse Relationship Between Nutritive Mass and Pecuniary Value at Eating Establishments."

This phenomenon helps explain the willingness of millions of consumers to spend big bucks for "lite" products — including the presumably pure bottled water often imbibed by sophisticated aspartame sufferers. Yet even this selection attained a degree of dubious notoriety during 1990 when benzene was found in batches of Perrier Water®.

"*I'll tell you how I lost 40 pounds. I began to read ingredients on labels.*"

Cartoon 4-1

©1987 Medical Tribune.
Reproduced with permission.

Weighty Matters

Obesity remains a favorite target for cartoon satirists. Variations of the I-eat-like-a bird (i.e., vulture) theme are a specialty. Example: a plump low-calorie devotee tells her dining companion, "Oh, I can eat any diet foods. In fact, I don't even bother listing sugar-free desserts in my diet diary."

Cartoonists are likely to take special delight in the phenomenon of "paradoxic weight gain" I described in previous publications (pp. 291-293). This refers to using aspartame products in order to lose or "control" weight...only to end up gaining lots more. Conversely, such persons are then likely to lose weight by avoiding aspartame products. A related theme appeared in a Medical Tribune cartoon, as one thin woman was telling her stout friend how she lost 40 pounds. The secret? Reading labels. (Cartoon 4-1)

The "Non" Sense Phenomenon

Some psychologists regard the current widespread acceptance of "non" sense as a form of self-denial. It extends beyond the use of noncalorie beverages, nondairy spreads, noncaffeine colas, nonfat ice cream and nonalcoholic beer to more demanding forms of abstinence.

This attitude probably reflects the desperate attempt by many persons to gain a sense of control through self-denial. One New York psychologist suggested that it serves as both penance for prior shameful behavior, and a panacea for anticipating the feared future...whether social, economic or medical.

"Just Say NO!"

Mention should be made of the "Just Say NO!" slogan. It is currently being directed against smoking, drinking, sexuality (especially with the specter of AIDS), and even the "easy money" of so-called junk bonds. The same could apply to avoidance of aspartame products. An effective variation asserts, "The 'right to know' also entails the 'right to no.'"

Humorists and cartoonists availed themselves of this theme.

- Irma Bombeck hoped that some of the new food labels mandated by the FDA and USDA by May 1993 would state "NO!" Why? This could provide a clue to locating products still having a flavor (The Palm Beach Post December 27, 1991, p. D-3).
- Sidney Harris provided another scenario in the May 4, 1990 issue of Science. (Cartoon 4-2)

Non-Anonymous Aspartame "Victims"

There was a dramatic multiplication of "Anon" organizations during the 1980s. They include Alcoholics Anonymous, Overeaters Anonymous, Gamblers Anonymous, Narcotics Anonymous, and Workaholics Anonymous.

Most aspartame "victims," however, don't seek anonymity. They want to shout their warnings from the rooftops, and via radio microphones and television cameras.

"IT'S A GOOD THING I BUILT THIS PLACE WHEN I DID. THE NEW ZONING LAW PREVENTS ANY HOUSE FROM GOING OVER 75,000 CALORIES."

Cartoon 4-2

5

THE WORLD'S FIRST "ASPARTAMOLOGIST"

Atlanta, 1986. I was scheduled to address the Section on Medicine of the Southern Medical Association about reactions to aspartame products.

A session preceding my talk dealt with disorders of the esophagus. A gastroenterologist joked about being an "esophagologist." The thought then struck: "Why, I'm probably the world's first aspartamologist!" A doctor sitting next to me kept looking in my direction, perplexed over my recurrent grin.

Another comment by the same speaker also hit a responsive chord. It concerned one of Sir William Osler's favorite themes. This great physician-teacher sought to remind doctors of their duty to educate the masses against taking medicine casually. I reflected, "Well, I guess it's now <u>my</u> function to educate the masses against taking aspartame products casually."

Cartoon 5-1

Another Version of the Emperor Clothes Syndrome

One rebuttal was repeatedly hurled at the first aspartamologist. It focused on a number of "negative" reports presumably disproving the notion that aspartame ingestion had any relationship to headaches, convulsions, altered behavior or other problems. In nearly every instance, however, the "scientific" study — almost always funded by some corporate interest — involved giving pure aspartame in a capsule or as a freshly-prepared cold drink rather than the incriminated product obtained directly from a market shelf.

This generated another scenario. I would ask or write the principal investigator a simple question: "Why didn't you just give these patients the same product that they alleged had caused the problem, and use a comparable product not containing aspartame as your control?" The response was predictable: a blank stare at the meeting, or failure to receive a written response to my letter.

Having gone through this experience several times, I can sympathize with the boy in that story about the emperor being duped into going naked. As you recall, the lad was reprimanded for uttering, "But he isn't wearing any clothes!" (See Chapter 44 for another variation of this theme.)

Another Doubting Thomas

My ears also go on high alert in two other situations. The first entails a physician, medical investigator or other health professional supported by the billion-dollar aspartame industry who proclaims total objectivity in conducting the foregoing "scientific" studies. The second involves insistence that he or she would never become involved in such a project "just for the buck." My skepticism over both remarks triggered a sense of kinship with "the doubting Thomases."

> I searched high and low for the identity of this presumed clan...but to no avail. The answer finally surfaced in Crowell's Handbook for Readers and Writers (1925), which made reference to the Apocryphal Acts of St. Thomas. Several legends generated the skeptic attitude. According to one, Christ allegedly sold him as a slave to an Indian prince then visiting Jerusalem when he refused to go to India as a missionary.

44

6

MIXING MEDICINE AND POLITICS

Medical perspectives and political views have a strange way of converging on seemingly mundane issues. They include additives and "fake" foods.

A Politician's Pavlov Response

My wife Carol is an elected official. She served four terms on the West Palm Beach City Commission...including the position of Mayor. Carol was subsequently elected to the Palm Beach County Commission, which she chaired for two years during her first term.

Carol's longstanding participation in many community projects as well as local, regional and state governmental activities that transcended political affiliations generated a large constituency. As a result, I often found myself rubbing shoulders with community leaders and high-level politicos.

One such person was State Representative R. Z. ("Sandy") Safley of Clearwater. He is a Republican; Carol is a Democrat. These party links, however, didn't influence their innovative approaches to mutual concerns such as fairness in homestead exemption, Florida's uncontrolled growth, and the State's outmoded workers' compensation system.

Carol mentioned my researches to Sandy on several occasions. This was prompted by his complaint of unexplained headaches, coupled with her noting his considerable intake of aspartame beverages. When I met Sandy during February 1989, he imparted the following experience.

> "Carol kept telling me about your studies on aspartame. So I decided to test the matter for myself. And she — or rather, you — were right! My headaches quickly disappeared! From that time on, I think of your name every time I hear about an aspartame product. What do you make of that?"

I smiled, but said nothing. My <u>actual</u> first thought now can be divulged: "Imagine! I cause a Pavlovian response in a politician!"

A cartoon in <u>The Miami Herald</u> (December 2, 1987, p. A-21) reinforced this humorous "Russian" connection. It depicted Mikhail Gorbachev promoting "Nutra Sweet USSR" on television. Our subsequent visit to Moscow (see 26), initiated by several proposed American-Russian joint ventures, suggested that this theme wasn't so farfetched after all.

Cartoon 6-1

The Dieting Politician

November 3, 1987...the day before general elections for many local, county, state and national offices.

Carol entered my study waving a copy of The Miami Herald. It featured a front-page article about Bob Martinez, then a Republican candidate for Governor of Florida. They had known each other for years, especially while serving simultaneously as the mayors of West Palm Beach and Tampa.

Bob campaigned vigorously. He also lost 25 pounds on a diet consisting of honey-coated cashews, MacDonald's hamburgers, coffee and diet colas. Reporters dubbed it "The Bob Martinez Weight Loss Plan."

Cartoon 6-2

Carol then related a spontaneous observation only two day previously by Betty, her perceptive mother. "Martinez doesn't look well, especially his remote smile."

I considered including a line concerning my researches on aspartame products in our letter of congratulations to Bob, but opted not to mix medicine and politics at that point. In the face of some subsequent gubernatorial decisions, however, I'm sorry I didn't.

Another Political Plug

The nation's eyes focused upon activities at Madison Square Garden during July, 1992 as the National Democratic Convention was being held. Its presidential nominee, Governor Bill Clinton, decided to check the platform in preparation for his July 16 acceptance speech. As the cameras recorded him taking a swig of a popular diet cola, the announcer quipped, "Boy, that's a great commercial plug!" As a registered Democrat, however, I had the same déjà vu about cautioning the candidate. But this time I did send a letter.

Political Sweethearts

Then there were the political contributions by persons associated with the aspartame industry. Some timely campaign donations made just after important Senate hearings, a matter of public record, raised suspicions about "sweetheart deals."

Prisoners' Delight

Carol had her first taste (pardon the pun) of Florida's major prisons during a personal survey of them in April 1989. She ate the same food as inmates, albeit not with them. Asking for "artificial sweetener," she was handed a huge container of a popular aspartame tabletop sweetener.

Carol's initial reaction was a combination of dismay and partial shock. She had heard me make repeated reference to severe psychiatric, psychologic, behavioral and personality changes experienced by hundreds of persons with reactions to aspartame products. In the present situation, it was difficult to resist lecturing the Department of Prisons about the potential aggravation of antisocial behavior by convicted felons after consuming large quantities of aspartame. Sensing that such comments would fall on deaf ears, Carol refrained.

Cartoon 6-3

More on Politics and Diet Sodas

Here's more advice to persons orbiting in the political arena: exercise <u>much</u> caution when ordering diet sodas in the company of certain executives. The same applies to individuals engaged in parapolitical pursuits, such as lobbyists for profit and nonprofit organizations.

This subject is nicely illustrated by an item that appeared in <u>The Palm Beach Post</u> (June 26, 1989, p. B-1). The Chairperson of the Florida Audubon Society was being flown to Tallahassee as a guest of Florida Sugar Cane League to witness the signing of an important bill. When asked about her preference for a beverage during the flight, she requested a diet soda. The League's startled representative responded, "Not on this plane!" The embarrassed environmentalist settled for water.

"The Cola Factor" in Medical Care Costs

Like it or not, political pressures and governmental regulations are having a major impact upon the cost of medical care. The ever-changing and declining maximum limits for physician fees and hospital services being set by Medicare bureaucrats also have created paranoia among doctors and hospitals. If you don't believe me, just allude to this subject the next time you meet a doctor or hospital administrator...and allow <u>plenty</u> of time for the response.

Faced with the quandary of how much to charge now for an office visit or an operation, Dr. J. N. Brouillette described his resort to "the Coke factor" in the May 1991 edition of the <u>Journal of the Florida Medical Association</u> (pp. 317-318). It stemmed from this advice given by a senior physician during the late 1960s.

> "Son, I have a foolproof formula for calculating reasonable medical fees. I charge according to the price of a bottle of Coke. When I started practising medicine years ago, it was a nickel and I charged $5 for an office visit. Now that the price of a bottle of Coke is 15 cents, I charge $15."*

---

*©1991 *The Journal of the Florida Medical Association.* Reproduced with permission.

This formula proved quite acceptable and accurate over the next two decades. For example, an office visit was $60 when cola averaged 60 cents a bottle. But there was one apparent gross discrepancy: the cost of malpractice insurance. Dr. Brouillette was paying $36,000 a year in 1989 instead of the predicted $4,200!

After lots of analysis, my colleague realized that he had not included "the legal factor" in his price for surgery. This entailed the costs of high-tech equipment and the numerous tests that now constitute the practice of defensive medicine to avoid the charge of "wrongful diagnosis" or "delay of proper diagnosis."

Dr. Brouillette did suggest a viable solution: let the government subsidize the cola industry, and return the price of a bottle to 10 cents. (This presumably would include diet colas.) As a result, the cost of an appendectomy would decline to $125.

A Congressional Fifth Column?

The executive director of a major industry based in Washington, D.C. called to relay the following "anecdote"...his exact word.

While dining with a member of Congress, the matter of severe reactions to diet drinks containing aspartame happened to arise. Being familiar with my writings on the subject, this businessman proceeded to elaborate on some of the problems encountered clinically.

The Congressman listened attentively. He then commented, "Maybe that's what has been going wrong around here! Lots of those sodas are being drunk in my own office!"

The REALLY Important Spouse

July 26, 1991. I was asked for an interview about aspartame reactions on ABC Television's The Home Show three days later. I explained to the producer that arrangements had been made months earlier to attend a seminar near Waterville, Maine. "No problem," she stated. She explained the details in a subsequent call. "I'm having a satellite dish set up at City Hall. You're scheduled for 11 A.M."

Carol and I arrived in Waterville an hour earlier. We were escorted to the interview area by the Mayor's secretary. I then jokingly remarked that my wife probably would be regarded as "the more important spouse" as far as City Hall was concerned.

Carol answered a few discrete questions posed thereafter by the secretary. It became evident that my seeming jest had merit, especially since Carol once served as the Mayor of West Palm Beach. The secretary stated, "Mrs. Roberts, the Mayor won't be here for an hour. But please do wait. I'm sure he'll want to see you!"

Before we entered City Hall, Carol informed me that she planned to leave the Council Chamber during the live interview "so that you won't be distracted by my presence." She kept her word, and visited with the Mayor of Waterville. The two happened to have much in common in "this small world." In fact, Carol cast the decisive vote to hire a friend of the Mayor as the Manager of West Palm Beach.

Possible Insights into a First Family Affliction

The public and the medical profession were enthralled when President George Bush and his wife developed an overactive thyroid — also known as primary hyperthyroidism and Graves disease. Speculation about some pernicious environmental factor mounted because their dog had developed lupus...another autoimmune disorder. However, public health sleuths were unable to uncover any noxious substance in their several homes. Most hands therefore concluded that the First Family's thyroid problem was coincidental. (Although Graves disease tends to run in relatives, George and Barbara Bush are not genetically related.)

This "coincidence" intrigued Yours Truly, an endocrinologist, for reasons cited below as well as several patient encounters. The question kept haunting me, "Could aspartame products have contributed to, or precipitated, Graves disease in the President and his wife?"

I wrote a doctor-to-doctor letter to the President's physician. After all, if the First Couple didn't use such products, the matter would require no further pursuit. My communication evoked a totally evasive answer that caused my antennas to stretch out further.

I experienced a Eureka phenomenon (see 28) several months later. The occasion was a two-day seminar on aspartame reactions held at the University of North Texas. It was ably coordinated by Professor Janet Smith (see Figure 27-1). She had called me one year previously about her own affliction with Graves disease. It developed after she began consuming large amounts of diet sodas and other aspartame products in order to attain "a mean fit look" at the urging of peers in her other professional capacity as an aerobics instructor (see 24).

Jan had enjoyed excellent health until then. The combined effects of an overactive thyroid and other symptoms attributable to aspartame products on this 35-year-old woman were devastating. They included severe fatigue, rapid heart action, headaches, visual disturbances, menstrual problems, "numbness and shooting pains in the arms and legs," and extreme swings in mood (including thoughts of suicide never previously experienced).

Jan balked when definitive treatment with radio-active iodine was recommended, particularly since no effort had been made to identify an environmental cause. She doggedly refused to ascribe the disease to "stress." Realizing that the only significant changes had been the "slender trap" from using aspartame products, she promptly avoided them. Her improvement was dramatic ...with normalization of all thyroid tests within three months. No recurrence occurred over the ensuing year in spite of a full academic load, instructing aerobics classes, and rearing three children.

What triggered the Eureka response? Her step-sister, aged 39, subsequently developed Graves disease! As in the case of the Bush couple, the two were not related since Jan had been adopted.

The step-sister, an insulin-dependent diabetic, began using aspartame products to avoid sugar. Shortly after starting their consumption, her blood sugar readings became unpredictable. There also was loss of urinary bladder control.

On Jan's advice, she avoided all aspartame products. The hyperthyroidism improved dramatically, and control of her diabetes and bladder function returned. The latter responses are consistent with my repeated observation that aspartame products can aggravate diabetes mellitus and its complications (see Publications).

I shall omit a lengthy dissertation on the mechanisms by which aspartame products might have triggered Graves disease in these step-sisters, other patients, and President and Mrs. Bush. They encompass excessive caloric restriction, increased energy demands relating to exercise and other activities, and metabolic derangements induced by aspartame and its metabolites. The following items, however, <u>are</u> germane to the Bush family.

- I received information "from highly reliable sources" that President Bush and his wife frequently consumed aspartame beverages and used a popular aspartame tabletop sweetener.
- As public figures, both spouses were highly conscious of their weight, and undoubtedly watched or cut back on their caloric intake.
- The 66-year-old President required lots of energy for his much publicized athletic prowess.

It is of further interest that five other persons had contacted the Aspartame Consumer Safety Network (Dallas, Texas) after concluding that such products probably contributed to the onset of their Graves disease.

7

A MESSAGE FOR GRANDMOTHERS

Neil Rogers is a controversial Miami radio talk-show host. He interviewed me for <u>two hours</u> on the subject of aspartame reactions, largely motivated by the fact that he was an "aspartame victim."

Listeners bombarded me with questions during the "open mike" segment. One caller was a concerned Jewish grandmother. Continually apologizing for her limited scientific vocabulary, she offered "some old-fashioned" ideas about what constitutes proper food. It wasn't necessary because many sophisticated nutritionists share her views, capsuled in the prescription, "Eat what your ancestors once ate."

I had been reflecting for more than three decades on how far Americans have strayed from certain eating habits that probably contributed to their ancestors' survival. Some thoughts crystallized as this lady spoke. When she finished, I boldly stated, "Neil, I've <u>finally</u> figured out what's been undermining our society," Puzzled, he asked me to explain.

> "It began when grandmothers decided to remain silent rather than risk being ridiculed for their presumed old-fashioned ideas. But I'm now giving this message to the Jewish, Italian, Irish, Black and other grandmothers in your audience: 'Get out of the closet, and start raising the dickens with your children and grandchildren about what and how they are eating! You <u>must</u> transmit this precious wisdom before they learn about it the hard way — like as patients in doctors' offices, clinics, and hospitals.'"

The ensuing prolonged silence stunned me and the host. I had clearly touched an important societal nerve.

Cartoon 7-1

"Nobody Listens to Grandma"

Others have pondered this phenomenon as well. The related heading of a column by William Raspberry in The Miami Herald (September 28, 1990, p. A-17) succinctly summarized my message: "Better Listen to Grandma."

David Blankenhorn offered two categories as the basis for what has gone wrong in our society in a recent issue of Family Affairs, published by the Institute for American Values. First, "Reagan closed the bathrooms" — referring to the shortage of public amenities by budget cuts during the Reagan Administration. Second, "Nobody listens to grandma."

Blankenhorn averred that one can learn more about the American family by speaking to ten randomly chosen grandmothers than ten randomly chosen "family experts." The former are more inclined to give tell-it-like-it-is opinions relative to value judgments. They range from a sense of right and wrong to establishing responsible priorities for the "new realities" of contemporary life.

I would add one pertinent P.S. from the comic strip Cathy. The founder of "Granny Nannies" indicated that if necessity were the mother of invention, desperation must be the grandmother (The Palm Beach Post August 15, 1991).

Views on "Nutritional Paternalism"

A raging controversy confronted members of the California Legislature during July 1991. It concerned extending the state sales tax to "snacks." But the Board of Equalization faced a dynamite political issue after voting for this tax: "When does a non-taxed food become a taxed snack?" For example, where would one draw the line between full-size cake or pie and snack-size cake or pie?

Brad Sherman was appointed Chairman of the Board. This accountant and tax-law specialist reached two conclusions (The New York Times August 3, 1991, p. 7). First, his constituents should have selected Julia Child (see 24). Second, the Board would not attempt "nutritional paternalism." Sherman added, "Our job is not to determine what's good for you, but to determine what's a cookie, what's a cracker. If you want to know what's good to eat, ask your mother." I would like to give Brad my two cents of advice: change mother to grandmother.

8

SORRY, ONLY SUGAR

The First International Meeting On Dietary Phenylalanine and Brain Function was held in Washington, D.C. on May 8-10, 1987. My scientific exhibit on aspartame reactions commanded the attention of many interested professionals.

I searched in vain for a saccharin packet or tablet at the first luncheon, but found only sugar. The waiter explained, "The Chairman of this meeting insists that we serve only sugar — no saccharin and no aspartame!"

After a bit of reflection, I realized that I couldn't fault the organizers for this tangible evidence of both neutrality and conviction.

A Potential Hooker

The solution for avoiding aspartame by using sugar, however, no longer can be taken for granted. Some producers have adopted that familiar tactic: "If you can't beat them, join them!"

Which brings me back to "sugar only." It seems that some consumers don't consider granulated table sugar sufficiently sweet. Accordingly, one company jumped onto the corporate bandwagon by introducing a product aimed at filling this void: a blend of cane sugar (99%) "sweetened" with aspartame (1%).

9

EMERGENCY RATIONS

Exposure to my researches on aspartame reactions made Carol, my wife, a "true believer." She also was familiar with my extensive studies on reactive hypoglycemia ("low blood sugar attacks") and diabetes over the previous three decades. All carried this "bottom line" of advice for persons so afflicted: "Avoid sugar. Avoid aspartame. Avoid overeating. Take small snacks."

Carol used saccharin or saccharin-containing packets in place of sugar and aspartame. Although there was no medical indication for doing so, she felt it prudent to respect the adage, "Where there's smoke, there's fire."

Carol kept encountering a problem, however — the absence of saccharin when dining out, even at fancy restaurants. Furthermore, some establishments carried only an aspartame-containing tabletop sweetener ("the blue packet"). So she resolved the matter by carrying her own "emergency rations."

This tactic had a forceful impact on the managers of several restaurants Carol frequented. She subsequently found "the pink packet" stocked in abundance...even without asking.

10

INTERESTING HEADLINES AND COMMERCIALS

One cadre of newspaper and magazine superspecialists has long intrigued me. They are the headline writers mandated to stimulate reader interest by providing provocative titles. An even more specialized subset make "headlines" at major sports events via the "battle of the blimps."

Captioning feature articles about <u>my</u> studies on aspartame reactions probably presented a challenge for this select group. But most performed with gusto. A few examples:

- "Aspartame Not Such a Sweet Ride?" (<u>Chronicle-Telegram </u>of Elyria, Ohio, August 3, 1986)
- "Some Bitter News About Top Sweetener" (<u>Post Time USA </u>October 1986)
- "West Palm Doctor Sour on Sweetener" (<u>The Palm Beach Post </u>January 8, 1990)
- "FDA May be Sugarcoating Harmful Effects of Aspartame" (<u>Wednesday Journal</u> of Oak Park, Illinois, February 7, 1990)
- "Sickly Sweet?" (<u>The Listener</u> [London], October 18, 1990)
- "Doctor Takes on a Sweet Giant" (<u>Morning Sentinel</u> [Waterville, Maine], July 31, 1991)
- "A Sour View of a Sweet Subject" (<u>The Miami Herald</u> January 5, 1990)

<u>The Miami Herald</u> reinforced the last-listed feature with this editorialized subtitle: "The FDA says it's dandy for candy. But Dr. H. J. Roberts is even quicker to bicker." (It was clearly a twist on the Ogden Nash ditty, "Candy is dandy, but liquor is quicker.")

Another headline in the June 9, 1988 edition of <u>The Palm Beach Post</u>

caught my eye. It read, "NutraSweet Hits Again." The article announced that the FDA had approved the use of aspartame in six more food categories.

Perception Vs. Reality: "Coke"

The need for restraint asserted itself when I found myself incorrectly perceiving the word "coke" in the headlines of medical and other periodicals, largely due to my interest in diet colas. For example, the title of a feature in Medical Tribune (March 25, 1987, p. 18) read, "Did NIDA Throw Coke Research A Fund Curve?" It became evident that I had erroneously inferred the substance in question: cocaine rather than cola beverages.

> I should add that there is a historical basis for such "misperception." Coca Cola® originally did contain cocaine until Dr. Harvey W. Wiley took action to make it unlawful.

This phenomenon struck with double impact while reading an editorial in The Palm Beach Post (p. E-1) on February 16, 1990. It was captioned: "The Coke Challenge: Optimism Justified, But Cutting Demand Tough." It concerned a recent "drug summit" attended by President Bush and three Central American leaders. President Virgilio Barco of Colombia set forth the problem in these succinct terms: "The only law the narco-traffickers do not violate is the law of supply and demand."

Why the double impact? On two radio talk shows the previous evening, I had made reference to the quasi-addiction of many persons who consume large amounts of aspartame beverages. In fact, some of these patients emphasized that their withdrawal symptoms were worse after attempting to stop these diet drinks than nicotine or alcohol! (I discussed this phenomenon in previous publications.)

My advice for persons with severe reactions to aspartame products is the same as to those who abuse drugs: "Just say NO!" This may not be an easy matter, however, as many spouses have discovered.

> • A woman suffered severe attacks of headache, mood swings, eye pain, drowsiness, itching, dizziness, diarrhea and abdominal pain. Numerous studies were "negative." She finally suspected the cause to be her consumption of many aspartame products. The

headaches disappeared within three days after avoiding them — and subsequently the other complaints. She wrote, "I purged my cupboards and threw all aspartame products into the garbage...Unfortunately, my husband is addicted to diet cola, but I'm still working on him."

• "Dear Researcher (God bless you!),

"I still use aspartame because my friends tell me 'everyone gets poor vision, hearing, and memory loss' at my age (50)...My husband and I have gotten into heavy arguments over my memory loss, and my reduced confidence over the loss of these 3 abilities which has interfered greatly with my ability to do several kinds of work."

Perception Vs. Reality: Misleading "Natural" Ads and Commercials

Scores of aspartame reactors have expressed indignation over less-than-candid headlines used in the mass advertising of "natural ingredient" products. Some correspondents did so with dramatic flair.

• A registered nurse who had experienced violent reactions to aspartame wrote

"In a state of innocence, we are poisoning our bodies with a supposedly 'safe substance' that is advertised as natural. However, in truth, it is produced in a factory and is not extracted from natural substances such as 'milk and bananas' as advertised."

• A 43-year-old aspartame reactor expressed resentment over several particular PR pitches in commercials.

"I hate the radio ad, 'Mother will never know it's (aspartame) rather than sugar, Dear.' And the TV ad with the cow that implies it is natural. PHOOEY!"

- A 30-year-old woman worked for <u>two</u> radio stations. She suffered severe confusion, impaired memory, insomnia, dizziness, and depression with suicidal thoughts while using an aspartame tabletop sweetener. These symptoms dramatically ceased after stopping the product. She stated

"I become angry when I see commercials promoting aspartame. How harmless they make it all look. The All-American brown & white can and something about bananas."

More on Coke

Let me rectify another historical matter that concerns "Coke." Insufficient attention has been paid to a gentleman named Coke outside the sphere of legal scholars. Indeed, his efforts in England four centuries ago played an important role in the evolution of our Fourth Amendment, especially the right of privacy.

While Attorney General to King James I, Sir Edward Coke composed the Great Reports. These 13 volumes covered 500 legal rulings between 1572 and 1616. In his notes on <u>Semayne's Case </u>(Volume IV), Coke preserved the famous doctrine that "a man's home is his castle." (It actually reads, "The house of every one is to him as his castle and fortress, as well for his defense against injury and violence, as for his repose.") This case, later a cornerstone for the development of English law in the colonies, helped ignite the American Revolution.

By Gumball!

I don't know if an "Emmy"-type award was given by the PR industry to the Chiat/Day/Mojo Agency in Venice (California) for originating the little red gumball that introduced NutraSweet® to the world a decade ago. It certainly deserved one. The accompanying shots of slim athletes personifying health reinforced the image of wholesomeness. The prominent display of this red gumball on billboards at a variety of international athletic competitions did the same.

Other PR firms weren't as fortunate with their "ball games"...particularly "eight ball" ploys.

> • The Pabst Brewing Company marketed a malt liquor called Olde English 800®. Its ad slogan, principally directed to poor urban blacks, was "Eight-Ball, Anyone?"
> • A rap jingle promoting St. Ides Malt Liquor® represented one-upmanship. Its message: "Forget eight ball... eight-ball aside for a stronger malt liquor."

The roof caved in, however, because "eight ball" is an illegal drug-market term for one-eighth ounce of crack cocaine. The inference from this argot was crystal-clear: one could get a cheaper high from these beverages (The Miami Herald July 5, 1991, p. A-12).

Let there be no confusion. I am not equating aspartame's red gumball and "eight ball." On the other hand, aspartame was developed as a drug intended for treating peptic ulcer (p.176). Furthermore, aspartame products can have profound pharmacological actions (especially on brain function) and lead to severe withdrawal symptoms, as noted above. My previous publications contain further details.

11

CORPORATE "ENEMY #1": IT'S A SMALL WORLD

Needless to say, my writings and media appearances concerning the subject of adverse reactions to aspartame products evoked more than passing interest by The NutraSweet Company. This was evidenced by the concurrent activities of its medical director (see 23) and consultants (see 36).

As emphasized in the introductory Disclaimer, my efforts and opinions were never meant as malice or disrespect to this Company or its representatives.

An Astonished Son

The perception of Yours Truly as a corporate ogre crystallized with extraordinary impact during a call from Stephen, my fourth son, on April 20, 1990. He had earned an M.B.A. degree from the prestigious Wharton School of Business. His skills in computer research and intelligence gathering were appreciated by members of the advertising industry, some of whom became close friends. One of Stephen's buddies related the following encounter, which surfaced in this conversation.

A PR firm used by The NutraSweet Company also sought the business of another company. Its representative presented the firm's credentials and policies at an open meeting. While explaining how it maximized effectiveness, he stressed a top priority: "Know your enemies."

A marketing expert and friend of Stephen happened to be in the audience. After the "Know your enemies" statement, one of the firm's executives remarked to him, "That means always being one step ahead of Dr. Roberts." The chap immediately recognized that this was a reference to Stephen's father!

Stephen ended the call with this sobering observation: "So, Dad, it seems you're now classified as 'Enemy #1.' Don't be surprised to learn that your mug shot has been posted on some corporate bulletin boards."

And that explains how I came to hold the new title of "Corporate Enemy #1" as well as "media terrorist" (see 33).

Cartoon 11-1

One Little Word

A remarkable tactic was repeatedly used in attempts to denigrate my published reports, medical lectures and interviews. Corporate representatives would reflexively assert, "It is virtually impossible for one product to be guilty of such a multitude of sins" (Morning Sentinel of Waterville, Maine, July 21, 1991, p.l). These "sins" — that is, complaints and illnesses — were listed in the Overview.

The word "virtually," of course, is critical. Anyone who doubts this need only study newspaper or television advertisements for the "virtual" safety of savings-and-loan institutions and companies selling securities. The same applies to "virtually" indefinite excellent performance of household appliances and used cars.

The low-key manner in which the "methyl ester" of aspartame is sidestepped or replaces any reference to free "methanol" or "methyl alcohol", its prompt breakdown product (see 17 and 26), is an art form that rivals the casual insertion of "virtually." When pushed further on this matter, corporate spokespersons offer several stock answers.

- "Sure, methanol can cause blindness if you drink something like white lightning."
- "A glass of tomato juice contains six times more methanol than a glass of soda sweetened with aspartame."
- "Why, you'd have to consume from 700 to 1,700 cans of a soft drink containing aspartame at one sitting to reach toxic levels!"

Rather than repeating the considerable evidence to rebut these replies (see Publications), I sometimes rested my case with the fact that free methyl alcohol — a poison — is rarely found in nature.

Allies

Being the "majority of one" challenging a huge industry often gets lonely. Many calls and considerable correspondence from aspartame sufferers bolstered the morale of "the first aspartamologist" (see 5). This letter is typical.

> "I just finished reading your book, <u>Aspartame (NutraSweet*):Is It Safe?</u> I am extremely grateful that you have taken such an interest in aspartame, and commend you on your considerable studies...It is a relief to know that I am not alone in my experience with this food hazard. Aspartame, with its powerful backers and medical ignorance, has the potential for being the health downfall of the world's population. It is a real tragedy that so few know this."

Consumer groups and individual aspartame reactors even expressed concern about my personal safety under these circumstances. I replied that such concern was exaggerated since (1) my comments were of a professional nature, and (2) the representatives of this industry, who understandably disagreed with my views, were considered honorable individuals working for respected firms.

Ultimate Evidence

A two-day seminar on the safety of aspartame products was held at the University of North Texas during November 1991. It dispelled any personal doubts about my status as "Enemy #1" from the vantage point of at least one manufacturer.

The last speaker on this symposium held the title, "Manager of Public Relations, The NutraSweet Company." She began by venting extreme criticism of the meeting itself. She then attacked the audacity of the University in sponsoring it. Her words: "We're here to protest the way this symposium has been organized because we believe the University and its students have been misled." The audience was aghast at this hostile tirade.

This PR rep made no bones about her feeling — namely, it was <u>not</u> her pleasure to be there. She proceeded to read a 10-minute message from Dr. Robert Moser (see 17). It repeated the corporate "party line," coupled with <u>his</u> version of a GAO report (see 42). In a final act of defiance, this spokesperson stated (1) she would not answer questions, and (2) anyone with a query should direct it to her at company headquarters. She then left...notwithstanding the large block of time that had been allocated for her presentation and rebuttal.

Her diatribe also mentioned the term "corporate neutral" several times. I had deliberately used these words in many of my writings, addresses, radio discussions, and television appearances to emphasize the need for neutrality in properly reevaluating aspartame products. These words, however, did not appear in the release issued to members of the media at this seminar — presumably because they might provoke poignant questions.

There was unanimous wonderment in the audience. How could this company, which had launched one of the most effective promotional campaigns in history, have allowed such a gross PR blunder? After all, each of the speakers was a person of good will who had come at his or her own expense to participate in a constructive dialogue within the neutral setting of a university.

I anticipated that such inflammatory venom would intensify efforts by these "activists" to have aspartame products removed from the market, and to hold corporate "spin doctors" and "crisis coordinators" fully accountable for their statements. This proved to be the case. The following excerpts from a letter by Ann Topper of Sanger, Texas, addressed to the company representative illustrate the point.

> "The purpose of the verbose hostile attack on the type of forum chosen for a public awareness symposium proved to be merely a half-hearted effort to mask the company's self-serving internal agenda...

> "What better place than a university setting, where young, eager, open minds are still capable of thinking for themselves because they aren't mentally anesthetized to the degree that the older generations are from decades of both subtle and aggressive invasive bombardment by the advertisers of artificial food products?...

> "It appeared as if you deliberately lambasted the type of forum chosen in order to avoid having to answer to the serious inquiries into the safety of the product.

Well, if the public relations manager is not responsible for addressing the public's inquiries and communicating educational information on the company's products, then who is? Your feeble attempt at trying to make the forum a sham backfired in your face...

"Furthermore, the curt, snippy replies to audience inquiries were noticed and noted, as was the fact that you arrived just in time to deliver the company's self-serving denial blurb, then exited just as swiftly. I say shame on all of you for insulting our intelligence and for your blatant lack of respect and decency to participate in a responsible and meaningful manner."

Waving the Red Aspartame Flag

The fiancé of an acquaintance had met the Roberts on several occasions. Aware of my interest in aspartame, she mentioned that her brother was a chemist working for Monsanto, The NutraSweet Company's corporate parent.

We attended a 1991 Christmas party hosted by a close friend. Shortly after entering, the fiancé approached me with a strange and aggressive gleam in her eye. "I'd like to introduce you to my brother. You remember? He's a Monsanto chemist."

The chap obviously had been primed concerning my reservations about the safety of aspartame products. Refusing to initiate a heated discussion in this social context, I stated, "It's a big subject, and no one knows all the answers."

The chemist cordially approached me later as we were leaving. He confessed to having made the decision to join another firm.

12

RIGHT SIZE, RIGHT WEIGHT

I mentioned Betty, my remarkable 85-year-old mother-in-law, in Chapter 6. One of her unique attributes is the ability to "tell it like it is" in a way that doesn't offend.

Betty had expressed several misgivings about my previous books. First, they were too long. Second, they weighed too much. Third, the print was too small. So she repeatedly pleaded: "Please, see that your next book isn't too heavy and has bigger print."

I presented Betty with a copy of my initial volume on aspartame reactions shortly after its release. She graciously expressed her appreciation, followed by a careful examination. I held my breath. The verdict: "Well, I think you finally have a book that's the right size and the right weight!"

Cartoon 12-1

I withheld one detail in the process. This was my small book on the subject, intended for the general reader. The scientific text already exceeded it at least fivefold.

13

TV & TOGETHERNESS

Over the years, Carol and I had appeared individually on a number of television programs. January 23, 1990 therefore proved a landmark day because we managed to appear on the <u>same</u> West Palm Beach television program. This separate-but-equal feat occurred during the Station WPTV (Channel 5) 5:30 P.M. news.

The first portion featured a dramatic breakthrough in Palm Beach County. Its Commission had forged ahead in providing affordable housing for more citizens. (Our area earned the unenviable reputation of being one of the most expensive markets in the country for prospective home purchasers.) The cameras focused on Commissioner Carol Roberts, whose political courage in spearheading a fair housing program received much praise.

The producer then shifted to my interview with his health reporter. She requested it after learning about my publications on the potential hazards of aspartame products (see Overview).

14

ROBERTS AND ROBERTSON...
AN ODD COUPLE

Reverend Marion G. "Pat" Robertson is famed for two reasons. He remains the star of The 700 Club. In 1988, he was a Republican candidate for President of the United States.

Robertson became interested in problems attributed to aspartame products because of his own reactions — specifically, severe mental confusion and memory loss. A request for viewers to share comparable experiences during 1986 generated more than 800 responses!

I received a call from Gailon Totheroh (see Figure 27-1), producer of The 700 Club, in December 1989. He had just read my article published in the peer-reviewed Journal of Applied Nutrition. On learning that a BOOK I had written on aspartame reactions was about to be released, he requested an interview at the Christian Broadcasting Network (CBN) studio in Virginia Beach. By coincidence, this appearance would be shown the very same day The Charles Press planned to release my book.

Reverend Robertson introduced the program. Once again, the station was inundated with scores of confirmatory "anecdotal" letters at his suggestion. I jokingly reflected on the number of political "foot soldiers" Robertson could muster merely by persisting with his anti-aspartame campaign.

Another producer for 700 Club On the Line, the nationwide sister "live" talk show, called me one month later. Would I agree to be interviewed and then answer calls for about one hour? I accepted the challenge. As predicted to this skeptical producer, every phone was ringing when the hour ended.

A disturbing incident occurred at the CBN mail room shortly thereafter. A package bomb addressed to Robertson exploded and injured one guard (The Miami Herald April 28, 1990, p. A-21). The minister surmised it was "part of a pattern of attacks against evangelical Christians." But an idea, admittedly paranoid, kept recurring: "Could someone have been more upset about Robertson's sermonizing against aspartame than over nonbelievers?"

An Odd Couple

These Roberts-Robertson scenarios had an element of irony because of our different religious persuasions. Robertson is a fundamentalist Christian. I am a Jew who prays daily. My nonmedical activities included founding the Jewish Federation of Palm Beach County and serving as president of the Jewish Community Day School of the Palm Beaches. Furthermore, one of my sons is currently a rabbinic student.

Even so, I felt a sense of spiritual exhilaration over the matter. Here were two persons from totally different backgrounds and professional pursuits whose interests had converged on a public health "mission" that transcended institutionized religious beliefs.

Television Coverage of a Symposium

The University of North Texas sponsored a symposium dealing with the safety of aspartame on November 7 and 8, 1991. These proceedings were covered from beginning to end by the Christian Broadcasting Network (CBN) under the supervision of Gailon Totheroh, mentioned earlier.

The crew put in extra time to cover the inflammatory tirade by the manager of public relations for The NutraSweet Company (see 11). Gailon interviewed me thereafter, with particular emphasis on her remarks.

I expressed dismay over such militant behavior within a neutral academic atmosphere that had been chosen precisely because it seemed conducive to constructive dialogue. Moreover, her absolute refusal to discuss <u>any</u> issues reflected poorly on this major corporation. I further stressed that (1) none of the speakers — representing physicians, consumers, the media, and the legal profession — claimed to have all the answers, and (2) they had come at their own expense both to impart information and to learn. This contrasted with the brief and presumably all-expense-paid "visit" of its representative.

A Lasting Impression

My wife, a recently re-elected Palm Beach County Commissioner, flew to Tallahassee on January 7, 1991 for several reasons. First, Carol had been selected as a member of the Task Force on Transportation by Governor-Elect Lawton Chiles in her capacity as Chairperson of the Tri-County Rail System that serves the Palm Beaches, Broward and Dade Counties. A meeting of this committee was scheduled that day. Second, Chiles, an old friend, would be inaugurated the next day.

The following conversation with two USAir stewardesses occurred en route.

Stewardess #1: "Would you like a cup of coffee, Ma'am?"

Carol: "Yes, thank you. And please bring me some Sweet N' Low®"

Stewardess #1 (handing her an aspartame-containing sweetener): "Here you are."

Carol: "No. That's _not_ what I asked for!"

Stewardess #1 "But aren't all sweeteners about the same?

Carol (sternly): "No, they aren't!"

Stewardess #1: "What difference does it make?"

Stewardess #2 (interrupting): "This one with aspartame can cause headaches, confusion, eye problems, memory loss, and depression."

Carol (impressed): "How do you know that?"

Stewardess #2: "Well, I heard a doctor discussing these problems on The 700 Club."

Carol: "Want to know something else?"

Stewardess #2: "Sure."

Carol (smiling): "You were listening to my husband. He wrote the book about aspartame reactions!"

15

THE BUCK STOPS HERE

The United States prides itself on being a "free enterprise" society. Its citizens have rejoiced in the potential for "upward mobility" through the determination and individualism personified by Horatio Alger. This theme also is embodied in the assertion, "Any child born in this country could become President."

I am not anti-business, as some might infer from comments about "self-serving corporate interests." I have worked hard as a solo primary care physician to meet payrolls, pay taxes, and support six children through and beyond college — always with the understanding that the rules for "entrepreneurs" in both business and the professions were fair and evenhanded.

Since entrepreneurs include corporate America, the question needs to be raised, "When is bigger and more powerful no longer fairer or better?"

- Justice Louis D. Brandeis stated in 1940, "Bigness is still the curse"...referring to his earlier statement in the 1934 collection of his papers The Curse of Bigness.
- This problem also was the basis for the book, America, Inc.: Who Owns and Operates the United States, by Morton Mintz and Jerry S. Cohen (Dell Publishing Co., Inc., New York, 1971).

What connection does this have with the present book? It would seem that BIG (spelled multi-billion dollar) industry is able to exert enormous influence on BIG government (read licensing and regulatory agencies) with regard to the marketing and promotion of certain products before their safety had been definitively confirmed. The case of the FDA will be discussed in Chapter 42.

The issue of "Sciencegate" also is troubling — in this instance, the potential influence wielded by corporate sponsors on investigators. Its pernicious nature has been evidenced in several headline stories wherein "Publish or Perish" became "Profit or Perish."

On Vested Interests

A legal definition of the term "vested," as in "vested interests," is "that which has become a complete and consummated right." In the field of business, it would be "cornering a large share of the market"...or even the Willie Sutton Business Principle ("gravitating to the money supply.")

> The subject of vests as garments has interested a few social scientists. Hein Schreuder of The Netherlands published a report in the Journal of Irreproducible Results (Volume 35, No. 1, pp.21-23) titled, "Suitable Research: On The Development Of A Positive Theory Of The Business Suit." He attempted to clarify the reason persons wear or don't wear three-piece suits — that is, with or without vests. An older colleague referred to such attire as "The Emperor's New Business Suit."

"Big Sugar"

In order to understand the attitudes of some vested interests to sugar substitutes, one must appreciate the position of the sugar industry. The latter happens to be a major enterprise in Palm Beach County, a fact not overlooked by our local media. Pat Crowley, a cartoonist for The Palm Beach Post, offered his rendition of the subject in its April 25, 1990 edition (p. E-3).

Cartoon 15-1

Some enterprising politicians have tried to amass voter brownie points by disassociating themselves from Big Sugar support. Such was the case with Barry Silver, a Boca Raton attorney, who challenged incumbent State Senator Don Childers during a Democratic primary race in 1990.

> Silver made a point of shunning contributions from this source. He stated, "We're not going to be sweet-talked by the sugar industry" (The Palm Beach Post July 2, 1990, p. B-1). To reinforce the matter, he offered sugar-free candy at political meetings. Even the Childers family couldn't refuse it. The incumbent said, "My wife says it tastes pretty good. Sugar-free candy is pretty expensive" (Sun Sentinel July 1, 1990, p. B-4).

16

ON SWEET TERMS OF ENDEARMENT

I have explained my concern over the massive consumption of both sugar and aspartame products in many publications. With reference to the real thing, someone once admonished Nature for letting sugar taste so good, but causing so much metabolic and medical difficulty. Nature guided us to safe, nutritious foods by making them sweet. We only ran into trouble when we separated the sweetness from the nutrients, so that refined sugar lacks the nutrients necessary to metabolize it.

I nevertheless have no qualms over "sweet" words of affection. In fact, attempted revisions of some "sweet talk" would be downright silly. Just consider "my saccharin pie" in place of "sweetie pie," "Equal® sixteen" instead of "sweet sixteen," and "my Sweet'N Low® heart" for "sweetheart." Such verbal abuse culprits could find themselves on the Willy Wonka most-wanted "sweet revenge" list of the Chocolate Manufacturers Association, especially during American Chocolate Week. They also would qualify on a comparable hit list sent to every florist and jeweler in the country, if one exists.

Cartoon 16-1

Some Personal Observations

We all have heard loved ones addressed as "Sweet" or "Honey." No such reference was more convincing then the utterance of "Sugar" by a dignified Southern matriarch whom I doctored many years. The term flowed from her lips as naturally as "mint julip."

The colloquialism extends to my immediate family — viz., I often call my wife "Sweetheart." Some fellow Rotarians and their wives were impressed by this tidbit when members had to confess the nicknames of their "Rotary Anns" at an annual dinner. (One chap reluctantly informed the group that he referred to his wife as "Barracuda.")

Yours Truly has made it a policy to maintain relative silence when Valentine's Day approaches. Since presents of candy and chocolate clearly would be inappropriate in my home, other gifts that convey affection are substituted. At times, this proved an expensive proposition. Carol didn't seem to mind, however, when I handed her a gift certificate that could be used at a famous jewelry shop on Worth Avenue. As far as she was concerned, most of the dictionary synonyms for this "sweet" act were just fine — like "agreeable," "pleasing," "affectionate," "loving" and "thoughtful."

Poetic Revisions

During earlier eras, poets and authors could be considered lucky in at least one respect. They were spared the onus of having their "sweet words" and "sweet dreams" undermined by certain reflexive thoughts that are evoked among contemporary readers. I refer to "calories," "cavities," and the need for a new wardrobe.

Even the titles of some current books must be chosen with care. Sweet Talk (Random House), a recent volume of short stories by Stephanie Vaughn, provides an example.

Had he known about aspartame reactions, I would venture that Jimmy Durante might have substituted "a diet drink" for "ink" in his hallmark offering about saying it with flowers or sweets, but never with ink.

A New Game

I even developed a novel game along these lines: substituting "saccharin" or "aspartame" for "sweet," "sweets" or "sweetness" in famous lines. "All love is sweet" from Shelley's <u>Prometheus Unbound</u> provides an illustration.

Let me therefore challenge any modern-day Bards of Avon to modify the following passages in this manner.

- "Sweets to the sweet: farewell!" (<u>Hamlet</u>)
- "Good night, good night! Parting is such sweet sorrow." (<u>Romeo and Juliet</u>)
- "It's a long way to Tipperary, to the sweetest girl I know!" (<u>Tipperary</u> by Harry Williams and Jack Judge)

Some amazing revelations surfaced while playing this game. I admit to gleaning unique insights, both humanistic and medical, from reading or rereading "sweet" poetic passages. Here are a few:

- "Whenever there is something sweet you will find something bitter, too."
  Petronius: <u>Satyricon</u> (60 A.D.)

- "Shun even a sweet which can grow bitter."
  Publilius Syrus: <u>Sententiae</u> (43 B.C.)

- "The sweets we wish for turn to loathed sours
  Even in the moment that we call them ours."
  Shakespeare: <u>The Rape of Lucrece</u>

- "A surfeit of the sweetest things
  The deepest loathing to the stomach brings."
  Shakespeare: <u>A Midsummer Night's Dream</u>

- "Sweet, sweet, sweet poison for the age's tooth."
  Shakespeare: <u>King John</u>

A Caveat

The contemporary male must tread cautiously in this realm. Ellen Goodman, an oracle on such matters, sounded a warning when the terms "sweetie" and "lovely lady" are directed to women who have reached the upper echelons of business, professions and politics. She warned that one man's chivalry could backfire as "another woman's chauvinism" (The Palm Beach Post September 28, 1990, p. E-3).

About the same time, Thomas H. Jukes (Department of Biophysics, University of California at Berkeley) became sufficiently exercised over several letters published in Science that he fired off one of his own. He wrote: "Worker ants are females, not he's...Remember, any worker can be a queen if she gets enough of the right food when she's young" (Science September 2, 1990, p. 1359). Considering the calorie-starved state of many contemporary teenagers afflicted with the fear-of-fat syndrome (see 24) who consume copious amounts of aspartame beverages, Jukes' observations had an even greater impact.

It may come to pass in my lifetime that women will express as much rage against female-oriented ads inferring that aspartame-containing soft drinks and foods are the Holy Grail of health and slimness as against comparable cigarettes ads. An ultimate in such gender outrage already has been leveled at tobacco companies by U.S. Surgeon General Antonia Novella. In a speech on October 3, 1990, she stated

> "We've walked a mile for a Camel, lived modern for L & M, been cool for Kools, and, oh yes, American women sure have come a long way, baby...Call it a case of the Virginia Slims woman catching up with Marlboro Man."

An Endearing Promise

It's no secret that women don't want to become premature widows or caregivers for their husbands. Cartoonists are attuned to these vibrations, especially relative to memory loss.

> One variation pictures a minister inserting this question during a wedding ceremony at the request of the health-conscious bride: "Do you, John, promise to eat only wholesome food, and to forsake all junk foods and aspartame products?"

The More-of-a-Sweetie Contest

One question seems to be foremost in the minds of some producers:

"How can we make things even sweeter?" This challenge stimulated major competition for sugar substitutes having high solubility in cold drinks and stability when used in baked goods. As noted earlier, aspartame is considered at least 180 times sweeter than table sugar or sucrose.

Here's the present state of affairs for the greater sweetness of some sugar substitutes...using table sugar as the standard. (Fructose is 1.2-2 times sweeter.) Not all of these sweeteners are approved or in use.

Cyclamate (sodium cyclohexyl-sulfamate) —	30 times
Licorice (glycyrrhizin) —	50 times
Abrusosides —	100 times
Aspartame —	180 - 200 times
Acesulfame K (Sunette®, Sweet One©) —	200 times
Stevioside —	300 times
Saccharin —	300 - 500 times
Sucralose —	300 - 600 times
Hernandulcin —	1,500 - 8,000 times
Neohesperidin —	1,500 times
Alitame —	2,000 times
Monellin —	2,500 times
Miraculin —	2,500 times

It is anybody's guess as to where the line ultimately will be drawn.

- A company now markets a product containing 99 percent sugar and one percent aspartame.
- Sunette®, heat-stable acesulfame-K, is being used to create an "endless" number of "sweeter dry mixes," such as a 50:50 blend with aspartame in a hot cocoa mix.
- A new compound (SC-45647), being developed by The NutraSweet Company, is 10,000 times sweeter than sucrose!
- Professor J. M. Tinti, Universite Claude Bernard (France), has been involved in NutraSweet's Sweetener 2000 project that involves families of substituted amino acids. They include tri- and tetra-substituted

guanidines and N-alkylated aspartic or glutamic acids. Some of these compounds allegedly have 20,000
times the sweetening potency of sucrose, while one
(sucronomic acid) is reported to be 200,000 times (!)
sweeter (Prepared Foods February 1992, p. 40).

There's a joker, however, in this enlarging deck of sweeteners —
namely, metabolic backfiring when the sense of taste has been conned.
For example, the excessive release of insulin in response to the expected
ingestion of real sugar (the cephalic phase of insulin secretion) can have
devastating consequences in terms of subsequent deprivation of glucose
("body sugar") within major organs, especially the brain.

The New Sweet Male

I eagerly plunged into the bowels of a feature written by Ellen
Goodman that appeared in the July 16, 1991 edition of The Miami Herald. It bore the headline, "The New Male: Tender, Sweet." My first
reaction: "Ahah! Must be some new insight about sweeteners!"

Fooled again by a tricky caption-writer! This article did not carry a
single word about the presumed subject. Rather, Ellen described her
reaction to the movie Regarding Henry.

> The central character, a stereotypical rich and ruthless
> lawyer, became "resurrected" after a "magic bullet"
> entered his brain. Without undergoing any con-
> sciousness-raising challenge, this attorney was trans-
> formed into a tender, family-loving and sweet-as-a-
> new-puppy guy. The nature of this change appeared
> validated by his hatred for the legal profession.

17

A TALE OF TWO DOCTORS

Through remarkable circumstances, I found myself engaged in a series of confrontations with Robert Moser, M.D., Vice-President for Medical Affairs of The NutraSweet Company. They occurred at medical meetings, in the print media, and on talk shows — both radio and television. Two physicians could not have been more polarized on a single issue...in this instance, the safety of aspartame-containing products for humans.

Bob and I had known one another since 1949. We were then the two (and only) medical residents on the same service at the Municipal Hospital in Washington, D.C. He was a Captain in the U.S. Army, which underwrote this and further residency-fellowship training. Bob subsequently assumed more conventional military obligations until leaving the Army. Thereafter, he consecutively became Editor of the <u>Journal of the American Medical Association</u>, and Executive Director of the American College of Physicians. For his "swan song," Bob opted for the foregoing position at The NutraSweet Company.

By contrast, Yours Truly planted his roots in West Palm Beach during December 1954. He has practiced there as a primary care doctor on a fee-for-service basis until the present time.

Who could have predicted that our interests would converge four decades later on the matter of a food additive?

Resurrection of a Quote

Bob authored a fine book in 1969 titled, <u>Diseases of Medical Progress</u>. I was impressed by this remarkable passage:

"Diseases of medical progress will be with us forevermore. They cannot be swept under the rug, either by clinician or drug producer. My own naivete in the world of commercial enterprise is revealed by my admission that I think a fine new drug will become known to the profession on the basis of its merit. I am embarrassed when this noble commodity is demeaned by merchandising techniques, however subtle or artful, better suited to less vital products, such as soap or *soda pop*."* (Italics supplied)

Mars-struck and Star Wars

The April 18, 1990 edition of the Journal of the American Medical Association carried an interesting item under "Miscellanea Medica" (p. 2028). It indicated that Dr. Robert Moser "is among members of a National Research Council committee that has told the National Aeronautics and Space Administration that overall plans for a lunar base and expeditions to Mars appear reasonable."

My reaction was twofold. First, I admired such remarkable involvement of a former colleague and military officer. Second, there was apprehension over the possible adverse effects on astronauts from the consumption of aspartame products during such a mission (a purely speculative inference on my part) — in particular, confusion, memory loss and visual problems.

The matters of space and space fiction unexpectedly arose 15 months later. Mary Nash Stoddard of the Aspartame Consumer Safety Network (see below and Chapter 36) checked with me about the time of my forthcoming interview on the July 29, 1991 program of ABC Television's Home Show. She made a passing remark during the conversation to the two chief adversaries in the film Star Wars. Nonplussed, I asked Mary to explain. "Oh, it's my reaction whenever I hear about another of your media sessions with Dr. Moser. It became reinforced by correspondence with several 747 pilots who told me of their severe reactions to aspartame products."

Sparring Via the Media

Certain sequences to my writings on the adverse effects of aspar-

---

* *From Moser, R.H.: Diseases of Medical Progress: A Study of Iatrogenic Diseases. 3rd edition, 1969, p. 819. Courtesy of Charles C Thomas, Publisher, Springfield, Illinois.*

tame products, or my discussions on radio and television, became as predictable as the setting sun. For example, Bob or his corporate associates would request equal — or Equal® — time (see 23). These confrontations contributed in no small measure to my being captioned "Corporate Enemy #1" (see 11), even though my opinions were generic and solely of a professional nature.

Our dialogues and literary sparring in the print media were unique. Bob sent rebuttal articles to the <u>Wednesday Journal</u> of Oak Park & River Forest (largely because of columns on this subject by Barbara Mullarkey, a nutrition writer), and <u>The Expert Witness Journal</u>. His letter to the EWJ was a response to my article published in its August 1989 issue.

Each successive attempt to discredit my remarks in varying degrees of abrasive language repeated the same corporate party line. As my data base and interest increased, I was tempted to remind Bob of the basic caveat taught to law students: "Don't ask an open question unless you know the answer!"

<u>Wednesday Journal Correspondence by Mary Nash Stoddard</u>

The following letter by Mary Nash Stoddard, President of the Aspartame Consumer Safety Network (ACSN), appeared in the May 9, 1990 issue of <u>Wednesday Journal</u>. I was not aware of its publication until receiving a copy, along with this nonprinted section.

> "Investigative reporter Barbara Alexander Mullarkey's comprehensive studies of <u>both</u> sides of the aspartame issue predate Moser's by more than five years! Perhaps Moser should have noted the facts contained in Mullarkey's excellent series of articles (in <u>Wednesday Journal</u>) before he signed on to defend the safety of a product which has such a disastrous track record in the 8 yrs. it has been on the market."

* * * * *

> We are amazed at the vicious attacks on author and dedicated clinician-researcher H. J. Roberts,

M.D., of West Palm Beach, Florida, and on brilliant Wednesday Journal writer Barbara Mullarkey by NutraSweet spokesperson Bob Moser (ONE VIEW, April 11).

Dr. Roberts' excellent, heavily-referenced book, Aspartame (NutraSweet*):Is It Safe?, raises many alarming questions about possible permanent damage and death from use of this substance. In actuality, Roberts treats patients while Moser fiercely "doctors" a faltering corporate image.

The "tirades" against all who question the safety of their chemical sweetener are completely unjustified. Moser grouped the thousands of people reporting adverse reactions to the FDA, the FAA and consumer groups such as ours as having "some axe to grind — unrelated to aspartame." We challenge Moser to name "the axe," rather than hiding behind false statements which cannot be sustained.

In his (ONE VIEW) of April 11, Moser states, "As most physicians, I knew very little about aspartame. It was not high on my priority of things I must know to... practice medicine." Yet at every opportunity, Moser admonishes aspartame victims to "discuss your health concerns regarding aspartame with a doctor," fully aware that most doctors know nothing about aspartame adverse reactions.

He constantly refers to "my antagonists." Those of us who publicly question the safety of aspartame have never mounted a personal attack against Moser as he has against us and our integrity. We have always clearly stated our campaign is against the FDA and their flawed approval of a neurotoxin. We want aspartame retested as a drug, which was how it was discovered....[see 28]

Our organization's position is that of seeking to call an irresponsible government agency, the FDA, into full public accountability. We have not attacked either corporations or individuals directly involved in product manufacture or promotions. ACSN's purpose: "Saving lives...through information and positive action."

Our challenge to Wednesday Journal readers: ask the next five people you see drinking a diet drink if they have had any of the reported symptoms — headaches, blurred vision, dizziness, gastrointestinal, etc. In our informal surveys, three out of five say, "Yes."*

The Expert Witness Journal (EWJ) Correspondence

The editor of EWJ requested my response to subsequent allegations by Dr. Moser. It appeared in the April 1990 issue (pp. 15-17), and is reproduced with permission. The editor's note (p. 20) concluded:

"It is our sincere hope that the medical concerns raised here and elsewhere over aspartame are answered to the satisfaction of all, preferably with convincing evidence of the product's safety. However, we must agree with Dr. Roberts and many others that these concerns should be addressed now, not 5 or 10 years from now when we may have a major public health hazard on our hands."

*   *   *   *   *

Harold D. Sewall, Esq.
Editor, The Expert Witness Journal
P. O. Box 590
Falmouth, MA 02541

Dear Mr. Sewall:

I am responding to the comments by Dr. Moser relative to my observations and concerns about products containing aspartame (NutraSweet®) that appeared in the Expert Witness Journal.

Since most readers of EWJ are trained in evaluating testimony, evidence and causation, a brief preface is offered for orientation.

(1) Dr. Moser believes that a good offense is the best defense. Accordingly, he uses terms such as "blatantly biased," "dangerously misleading," "false information," and "outrageous" (in another letter to EWJ). My position is one of seeking and honorably reporting the truth as best as I can without personal or corporate bias.

(2) Dr. Moser works for a billion-dollar company (a subsidiary of Monsanto). By contrast, I have received no salary or research funds from any vested interests in this area.

(3) I am a physician engaged in giving primary care and seeing difficult problems in consultation. To my knowledge, Dr. Moser has not involved himself in private medical practice for many years...and therefore cannot discuss these issues on the basis of first-hand encounters as a practicing doctor.

The following are "just the facts," supplemented by new and truly "alarming" information. Many of the details appear in my recent book, ASPARTAME (NUTRASWEET*):IS IT SAFE? (The Charles Press). I make special reference to epidemiology and the so-called signature effects (syndromes) encountered in persons having reactions to aspartame products for what I regard as "well-founded methodology." If Dr. Moser chooses to ignore these facts, or tailor the data to suit his own corporate purposes under the guise of "scientific" and "controlled" studies, that is his option.

- ITEM: "To say 'there were no extensive pre-marketing studies on humans before aspartame was licensed' is not true." I challenge Dr. Moser to provide me with a single report of extensive and detailed studies on the administration of aspartame products (especially in hot or stored drinks) to humans before the chemical was licensed in July 1981 — with special attention to side effects.

- ITEM: "There is an overwhelming body of evidence showing aspartame is safe for the general population." In effect, there has been an overwhelming and unprecedented number of reactions to aspartame products received by the FDA, the CDC, the manufacturer, several national consumer groups (e.g., Aspartame Consumers Safety Network), and interested physicians or investigators such as myself. The FDA already has logged over 5,000 reactions volunteered by consumers. In fact, more than 80 percent of the complaints about foods and additives it receives relate to aspartame products! As an indication of their serious nature, the FDA has reports on over 250 cases of convulsions. My own registry of more than 600 aspartame reactors contains over 100 persons with seizures.

- ITEM: "Prior to FDA approval of aspartame in 1981, more than 100 studies in animals and humans were conducted." Once again, I challenge Dr. Moser to provide me with data on human studies that had

sufficient numbers for proper epidemiologic evaluation meeting the standards expected by a qualified
statistician.

- ITEM: "The safety of aspartame has been well documented in numerous controlled studies..." The definition of "controlled" by Dr. Moser, especially in terms
of the "scientific studies" to which he alludes in persons having "headaches, dizziness, depression,
rashes, mouth and throat reactions, hyperactivity, extreme fatigue," are <u>divorced from reality</u>. Virtually
every corporate-sponsored "scientific" challenge involved a protocol wherein the subject was given pure
aspartame in a capsule or a freshly-prepared cold
drink. Stated differently, subjects did not drink
sweetened hot liquids, nor were they given <u>the same</u>
product (especially a beverage) obtained in a market,
and previously transported and stored for an unde-
termined period, perhaps in the heat. This is an important point because aspartame breaks down into a
number of other chemicals which also may have profound pharmacologic or toxic effects. For perspective, Israel and other countries not only warn consumers about exceeding a specific amount of aspartame, but also against the storage of such products in
heat and their use after a specific expiration date.

- ITEM: "Obese persons." In my opinion, persons
with obesity and other eating disorders should not
use aspartame products. The problem becomes even
more complex because of "paradoxic weight gain."
[see 24]

- ITEM: "Lactating women"..."pregnant women." In
my opinion, <u>no pregnant woman nor lactating
mother should risk exposing a fetus or infant to large
amounts of phenylalanine and methyl alcohol during</u>

<u>these critical phases of central nervous system devel-
opment</u>.

- ITEM: "Dr. Roberts' testimony as a 'expert witness' is
full of inaccuracies." I have <u>not</u> yet given a deposition nor have I testified as an expert witness in this
realm. In the event such services were to be requested
in a case that I regard as meritorious, I would consider serving in this capacity. [*Comment.* This situation remained unchanged as of May 1992.]

- ITEM: "Aspartame does not cause methanol poisoning, visual impairment, or blindness." About one-
fourth of the aspartame reactors in my series have
had severe disturbances of vision in one or both eyes;
blindness occurred in one or both eyes in 14. While it
is true that phenylalanine and aspartic acid also
might affect vision for several reasons, methanol is
certainly the most logical cause. I would point out
that I am referring to <u>free</u> methanol, which is rarely
found in nature, and comprises 10.9 percent of the
aspartame molecule.

I have reviewed the disinformation that Dr.
Moser continues to set forth regarding the purported
high concentrations of free methanol in fruit juices,
and address this matter in my book. Furthermore,
aspartame-containing beverages do not have the associated ethanol and acetaldehyde in fruit juices that
tend to offset the effects of small amounts.

- ITEM: "The scientific and regulatory literature..." In
effect, virtually all studies and reports on aspartame
have been done under either grant or contract with
the manufacturer or producers, or by organizations
they support. <u>To the best of my knowledge, neither
the FDA, the WHO, nor the regulatory agencies of</u>

the 80 countries that Dr. Moser continually cites have conducted independent animal and human studies to confirm the earlier reports, especially about toxicity and brain tumors. Most merely rubber-stamped the previous literature and bibliography.

• ITEM: "Diabetic persons..." As an endocrinologist, it has been my repeated experience that aspartame products can (1) aggravate diabetes mellitus, and (2) either intensify or simulate diabetic complications (especially neuropathy and retinopathy). My repeated requests of one diabetologist to explain the alleged "lack of significance" of striking blood sugar elevations in diabetics given aspartame, based on his own data, have never been answered. Yet, this is one of the "scientific" studies referred to by Dr. Moser.

• ITEM: "There is no substance known to man that can cause the diverse effects that Dr. Roberts has cited..." Yes, there is — products containing aspartame. If Dr. Moser doubts this, let him examine Table 7-2 in my book (p. 55) that lists complaints reported to the FDA by 3,326 aspartame complainants in an earlier study.

Dr. Moser made other misleading references to the editor of EWJ in a more lengthy previous letter.

• ITEM: Dr. Moser failed to indicate that the Acceptable Daily Intake (ADI) issued by the FDA does not pertain to human data, but represents a projection of animal studies based on lifetime intake.

• ITEM: Dr. Moser's analysis of the Government Accounting Office's two-year investigation of aspartame again makes no reference to the fact that more than half of the scientists surveyed were concerned about the neurologic reactions and other potential adverse effects of such products on children. In fact, 15 percent suggested a total ban. The Acting Comptroller General of the United States wrote to Senator Metzenbaum on June 18, 1987:

"We did not evaluate the scientific issues raised concerning the studies used for aspartame's approval or FDA's resolution of these issues, nor did we determine aspartame's safety. We do not have such scientific expertise. However, we did send a questionnaire to researchers to obtain the views on aspartame safety and information on aspartame's research."

- ITEM: The CDC [Centers for Disease Control] report on adverse reactions to aspartame products is repeatedly interpreted by Dr. Moser as if he read it while wearing corporate rose-colored glasses.

It is now my turn to ask Dr. Moser for his answer to "accurate, balanced, and reliable information" from a data base with unimpeachable scientific credentials: the National Cancer Institute SEER Program. I specifically refer to <u>the dramatic increase in the age-adjusted incidence rates of primary malignant brain tumors that has been recorded in all races and in both genders — beginning in 1984 or 1985, and persisting and increasing through to 1986 and 1987</u> (the last year for which complete statistics are available.) Furthermore, these increased incidence rates, using the 15-year statistics, also indicate striking increases <u>annually</u> from 1983 through to 1987 according to the Estimated Annual Percent Change (EAPC).

I have reviewed these data with a statistician who assures me that the National Cancer Institute regards them as a real statistical phenomenon. I do not believe they are explainable by improved diagnostic methods since good diagnostic scanning methods have been available for over a decade.

I would point out three facts to Dr. Moser that must be considered, and for which his reiteration that my "opinions should not be taken seriously" ought not be casually made.

- The high incidence of malignant brain tumors encountered in rat studies following the administration of aspartame during the 1970s [see 34], and for which the statutes of limitations were allowed to expire before the Delaney Clause to the 1958 Food Additives Amendment could be invoked.

- The refusal of a Public Board of Inquiry to approve the safety of aspartame relative to brain gliomas in treated rats "until additional studies are carried out using proper experimental designs."

- The consumption of aspartame products by more than 100,000,000 persons in the United States since 1981, and especially since July 1983 when aspartame beverages were approved. Remember: it took five years before thalidomide was recognized as a public health hazard.

Dr. Moser owes me and the readers of EWJ detailed and accurate answers to every one of the above items before the verdict is reached about who is "blatantly biased" concerning this possible "recipe for disaster."

Agreement on "Gratuitous Advice"

Dr. Robert Moser reviewed his multiple careers in The Pharos, journal of the honor medical society Alpha Omega Alpha to which I also had been elected in 1946. This article, "Serendipity Can Be Nudged," appeared in the Summer 1991 edition (pages 27-31).

I completely agreed with Bob on two matters — the appropriateness of his title, and this gratuitous advice he offered: "Do not be intimidated by something that is a little different, if it has great appeal. Life should be an adventure."* Who could doubt that my recent adventures in this unchartered "little different" subject held great appeal for me? They also were fraught with a degree of professional intimidation that few physicians would ever encounter.

> The public relations manager of The NutraSweet Company told the Morning Sentinel of Waterville (Maine): "We totally disagree with Dr. Roberts. His claims are not based on scientific evidence. They're outlandish and very emotional. We refute him. This is just the banner he's waving" (July 31, 1991, p. 2).

Lousy PR

My wife repeatedly heard an old adage from her wise mother: "If you can't say something nice about a person, keep quiet!" The implication was clear — reflexive badmouthing ultimately tends to backfire.

Another validation of its merit surfaced during a symposium on aspartame safety held at the University of North Texas (see 14). Professor Jan Smith did a remarkable job coordinating the seminar. Wishing to be fair, she invited the manager of public relations for The NutraSweet Company to give the final presentation...in effect affording her the last word. But Professor Smith made one specific condition: Dr. Robert Moser was not to replace her.

I was not aware of this matter until the conference ended. Jan offered this explanation to the PR person.

> "I have no objection if Dr. Moser comes with you, but not as the Company's main representative. I heard Dr. Moser make several remarks that were totally uncalled for, particularly his insulting statements about Dr. Roberts on a recent Home Show program that was shown nationally. I refuse to have my students exposed to denigration of anyone, especially a caring physician like Dr. Roberts."

---

* © 1991 Alpha Omega Alpha Honor Medical Society. Reproduced with permission.

18

A WOMAN'S BEFORE-AND-AFTER PREROGATIVE

I want to emphasize or clarify two matters. First, Raquel Welch is a beautiful actress whose natural endowments should be appreciated...even by medical practitioners. Second, a women is entitled to change her mind both about appearances and diet.

Leading magazines printed a multicolored ad during 1986 that featured a picture of Miss Welch. She held a glass with the imprint "Crystal Light." Its heading proclaimed: "ABOUT YOUR DIET SOFT DRINK. NOW TRY MINE." The bottom of the promo asserted: "I BELIEVE IN CRYSTAL LIGHT CAUSE I BELIEVE IN ME."

I experienced a strange déjà vu, but couldn't account for it. This puzzle haunted me for months. The answer then struck while arranging slides for a lecture on the hazards of extreme caloric restriction. I had resurrected a slide made decades earlier showing two photos of a curvaceous actress who posed for "WATE-ON." This product carried the promise: "Helps Put Pounds and Inches On Skinny Figures When Underweight Is Diagnosed As Due To Poor Eating Habits." The title read, "TRUE BEAUTY INCLUDES A FULL FIGURE."

Lo and behold! This ad also carried a testimonial. "I Can't Afford To Be Skinny...says Glamorous Actress RAQUEL WELCH."

19

CONFUSION, MEMORY LOSS AND "OLD AGE"

My studies on persons with adverse reactions to aspartame-containing products indicate that nearly *one-third* experience significant confusion and memory loss. Some examples:

- "I had difficulty remembering who I was talking to while on the phone." (Prominent executive in my community)
- "I couldn't think straight." (Building inspector)
- "I lost my photographic memory." (47-year-old woman)
- "I had trouble sequencing assignments after I started to drink two cans of diet cola daily." (34-year-old teacher)
- "I was frustrated because of inability to pull facts out of my mind, and to put them together in my mind. Some things seemed to be lost." (55-year-old home-maker)

The severe confusion and memory loss that affected a 45-year-old woman while consuming aspartame products will be described in Chapter 40. She wrote, "If something isn't done to warn about the dangers of using aspartame, we're going to have a population of mind-less zombies!"

Two additional points deserve emphasis. First, the vast majority of such sufferers were under 60 years. Second, "older" persons with existing memory problems ought to avoid aspartame products because of the added risks and stigma. Many patients in both categories promptly related to Cartoon 19-1.

Cartoon 19-1

Some good-natured patients and friends even relayed their pertinent jokes and cartoons.

- One patient acceptingly joked, "If you are losing your memory, forget about it."

- Another sent this classic version (original source unknown)

  "The minister called the other day, and told me I should be thinking about the hereafter. I told him that I do so all the time. No matter where I am — in the parlor, upstairs, in the kitchen, or down in the basement — I ask myself: 'What am I here after?'"

- A nationally known nutritionist concocted this variation.

  Patient: "Doctor, I'm suffering from memory loss."
  Doctor: "How long have you been suffering?"
  Patient: "Suffering from what?"

"Rejuvenation"

This practising doctor found a repeatedly encountered experience to be uniquely gratifying. It involved patients who had been profoundly concerned about the possibility of having "early Alzheimer's disease." The dreaded confusion then vanished after they stopped using aspartame products, as did their terrible associated anxiety.

Since many persons over 60 depend on being able to drive a car safely for their social and/or financial independence, such reassurance that the computer being supported by their necks was properly functioning generated special appreciation. Their continuation of animated constructive activity attested to the declaration by Pindar (c.500 B.C.): "We don't crave immortality, but we must reach out to the limits of what is possible for mankind."

Humorists, of course, have offered other newsworthy recommendations for assisting an "unreliable memory."

> Erma Bombeck emphasized the need to decide what information <u>not</u> to remember (<u>The Palm Beach Post</u> November 20, 1990, p. D-3). Such useless tidbits included clearing one's brain of the names of friends' divorced spouses, the duration of labor with each child, and in Erma's case, her age. Concerning the latter, she figured that if other persons regarded this detail to be so important, <u>they</u> should keep track of it.

On "Middle Age" and "Old Age"

My dealings with "seasoned" patients often provoked discussions about "middle age" and "old age"...including the definitions of these terms. One should be leery, however, about venturing into this quicksand.

> Ogden Nash regarded middle age as the time when a person has met so many others that each new acquaintance reminds him of someone else.

There's another reason for caution. Most chronologic parameters used in the past are being modified, or even discarded, as our population "matures." I <u>commonly</u> encounter persons in their 70s and 80s who exhibit more sparkle and mental agility than some thirtysomethings. In fact, "old age" or "the third third" is subclassified nowadays as "the young old" (perhaps more accurately the "old young"), "the middle old," and the "old old".

I have found it helpful to provide patients and their relatives with perspectives concerning these vexing issues that transcend simple chronology. Here are a few favorite sayings that usually hit the mark.

- "Age is a matter of living, not years."
- "We all have to get older, but not old."
- "To stay youthful, remain useful."
- "Age is important...but only if you are wine or

cheese."
- "Age is more a function of gray matter than gray hair."
- "We do not count a man's years until he has nothing else to count." (Ralph Waldo Emerson)

Generation "Passages"

Persons over 40 who no longer qualify as "The Pepsi Generation" often join "The Diet Soda Generation." Unfortunately, they risk monumental confrontations when aspartame-related problems ensue. Psychiatrists and psychologists are likely to see them for presumed "midlife crises" involving sexuality, memory problems, business crises, etc.

As aspartame products continue to inundate the world, professionals in other countries must brace themselves for comparable encounters.
- Germans refer to "midlife crisis" as schluspanik (the "fear of closing gates"). The French use the term "après du midi."
- The Japanese tend to adopt an attitude of "resistance" in their perception of "aging".

"Middle Age"

I never knew where my open-ended researches on aspartame reactions would lead me. Such was the case when it became necessary to gather more statistics on brain tumors in the face of their apparent increased incidence following the introduction of aspartame products (see 17 and 34).

While conversing with a National Cancer Institute statistician, I pointed out that brain cancer previously had occurred chiefly among "middle-aged men." She concurred. I then shifted gears.

H.J.R.:	"You know, I find it's getting harder to figure out just where middle age begins and ends."
Statistician:	"The same with me! It seems the age limit keeps rising all the time."
H.J.R.:	"For example, do I sound middle-aged to you?"
Statistician:	"Hardly!"
H.J.R.:	"What would you say if I told you I'm over 66."

Statistician:	"It proves the point."
H.J.R.:	"Thanks. You just made my day!"

Lessons From a Retired Nurse

I received a completed questionnaire from a 79-year-old retired nurse and teacher. Her first reaction to an aspartame soda had occurred at the age of 76. The symptoms included double vision, severe headache, depression (for the first time), dizziness, unsteadiness, drowsiness, confusion, and marked memory loss.

Some of her experiences and perceptive observations warrant repetition.

- "I had amnesia for about 20 minutes. Thought I might have had a cerebral hemorrhage, so went to the doctor. With medication for a wrong diagnosis (psychomotor epilepsy), I was ill all summer."

- "I assumed the posture of many people in nursing homes. I sat with my hands in my lap, eyes closed. I had double vision with or without glasses, so I did not wear them. I just sat."

- "My first experience with an artificial sweetener was on a long plane trip when good drinking water was not available. The hostesses had the attitude that I was 'odd' to be asking for water in large quantities. So I finally succumbed to their pressure and drank 1-calorie pop. I dozed during the trip. When I woke up, the people in front of me were hanging upside down from the ceiling of the plane! With great effort, and by blinking frequently, I finally opened my eyes; the people were all right side up. At that time, I thought it was an eye problem."

- "I saw another doctor who took me off all medication. I gradually recovered and could drive a car after 3 weeks."

"Forever Youthful" Reflections

Please don't mistake my emphasis, Dear Reader. Avoiding aspartame products alone will not reverse the clock. And neither will the use of scores of anti-aging items championed by the health and cosmetic industries. In fact, someone astutely observed that they are unique in at least one respect: their representatives and speakers tend to brag about being older than they actually are.

Conscience has forced others to provide similar qualifiers. Dr. Arnold Lorand, an Austrian physician, prefaced his 1910 book, <u>Old Age Deferred</u> (F. A. Davis Company, Philadelphia), with the statement that no reader should assume it could transform elderly persons into "sprightly adolescents."

20

"PRODIGIOUS" PERSPECTIVES

The term "prodigious" is defined in Webster's <u>New Collegiate Dictionary</u> as "portentous," "vast," "huge," and "extraordinary in bulk, quantity, or degree."

I actually made a conscious effort to avoid repeating this word when discussing the magnitude of aspartame consumption, both by individuals and on a national basis. But I found myself drawn back to it as if by a giant magnet.

Two thoughts arose as I reflected on the reasons. One was that familiar justification parents hear: "But everybody's doing it!" (This encompassed the phenomenon of lemmings drowning at sea because "everybody's doing it.") The second was a sentiment expressed by Mae West, "Too much of a good thing is wonderful."

A few facts illustrate the scale of this "marketing miracle of the 1980s." In a sense, they attest to Oscar Wilde's cynical remark, "Progress is the realization of Utopias."

- According to ads appearing in June 1991, aspartame was present in over 4,000 products being used by more than 200 million persons. The current figure is 4,200.
- By 1985, the American public already had consumed an amount of aspartame equivalent to 1.66 <u>billion</u> pounds of sugar.
- The dramatically increased sales of the two leading brands of aspartame-containing colas during 1986 elevated their ranks to positions three and four among all carbonated beverages consumed in the United States.
- The 1989 sales for sugar substitutes were estimated at $1,090,000,000 (<u>Prepared Foods</u> September 1990, p. 58). A 9.0 percent growth rate of such sales through

1995 was forecast.
- Someone calculated that the consumption of aspartame by Americans during 1987 equalled the weight of 30 blue whales. (A mature blue whale weighs 150 tons.)
- Low-calorie, sugar-free products constitute a massive infiltration of foods and beverages in the United States and abroad. They are consumed by 54 percent of adults in the United States, 47 percent of adults in Germany, and 35 percent of adults in the United Kingdom (Food Product Design April 1992, p.17).

Well, we just beat last year's production!

Cartoon 20-1

For readers who want "just the facts," preferably in tabulated form, I include the following table. It appeared in <u>Sugar and Sweetener: Situation and Outlook</u>, published in June 1988 by the Economic Research Service of the United States Department of Agriculture (USDA).

U.S. per capita consumption of caloric and low-calorie sweeteners.

Calendar Year	Refined sugar	Corn sweetener[1] HFCS	Glucose	Dextrose	Total[2]	Total of all caloric sweeteners[2,3]	Saccharin	Aspartame	Total of all low calorie sweeteners[4]	Total of all sweeteners	Population
			POUNDS PER CAPITA								MILLIONS
1970	101.8	0.7	14.0	4.6	19.3	122.6	5.8	0.0	5.8	128.4	205.1
1971	102.1	0.9	14.9	5.0	20.8	124.3	5.1	0.0	5.1	129.4	207.6
1972	102.3	1.3	15.4	4.4	21.1	124.9	5.1	0.0	5.1	130.0	209.9
1973	100.8	2.1	16.5	4.8	23.4	125.6	5.1	0.0	5.1	130.7	211.9
1974	95.7	3.0	17.2	4.9	25.1	121.9	5.9	0.0	5.9	127.8	213.9
1975	89.2	5.0	17.5	5.0	27.5	118.1	6.1	0.0	6.1	124.2	216.0
1976	93.4	7.2	17.5	5.0	29.7	124.4	6.1	0.0	6.1	130.5	218.0
1977	94.2	9.5	17.6	4.1	31.2	126.8	6.6	0.0	6.6	133.4	220.2
1978	91.4	12.1	17.8	2.8	33.7	126.6	6.9	0.0	6.9	133.5	222.6
1979	89.3	14.9	17.9	3.6	36.3	127.1	7.3	0.0	7.3	134.4	225.0
1980	83.6	19.1	17.6	3.5	40.2	125.1	7.7	0.0	7.7	132.8	227.7
1981	79.4	23.2	17.8	3.5	44.5	125.1	8.0	0.2	8.2	133.3	230.0
1982	73.7	26.7	18.0	3.5	48.2	123.2	8.4	1.0	9.4	132.6	232.3
1983	71.1	30.7	18.0	3.5	52.2	124.6	9.5	3.5	13.0	137.6	234.5
1984	67.4	36.3	18.0	3.5	57.8	126.6	10.0	5.8	15.8	142.5	237.0
1985	63.0	45.0	18.0	3.5	66.5	130.9	6.0	12.0	18.0	142.4	239.3
1986	60.2	45.6	28.0	3.5	67.1	128.7	5.5	13.0	18.5	148.9	241.6
1987	62.2	47.3	18.0	3.5	68.8	132.4	5.5	13.5	19.0	147.2	243.9
1988[5]	62.6	48.0	17.9	3.6	69.5	133.5	6.0	14.0	20.0	152.4	246.1

[1] *Dry basis*
[2] *May not precisely add to total of individual items because of rounding.*
[3] *Includes honey and edible syrups.*
[4] *Sugar sweetness equivalent. Assumes saccharin is 300 time as sweet and aspartame is 200 times as sweet as sugar.*
[5] *Forecast*

Two aspects of this revolution created by aspartame illustrate the point.

- Worldwide, more than 100 frozen novelties, 50 frozen yogurts, and 80 frozen dairy desserts now contain this sweetener (Prepared Foods May 1991, p. 123).
- Full-page ads for a "bulk pack" of aspartame appeared in Food Arts during 1991. (This is a journal for professional cooks.) The promos suggested its use "in many of your favorite recipes," and "to lighten popular drinks you make by the batch."

The massive intake of products containing aspartame by some individuals with reactions to them has been noteworthy.

- A 29-year-old patient with aspartame-induced seizures drank 18 (!) twelve-ounce cans of an aspartame cola beverage daily for one year.
- A 35-year-old pilot with convulsions, severe drowsiness, slurred speech, severe depression, irritability and intense headaches had been consuming these amounts every day:

  - Two to four 16-ounce bottles of aspartame soft drinks
  - One 2-liter bottle of aspartame soft drinks
  - Up to 12 packets of an aspartame tabletop sweetener in hot tea
  - One or two regular glasses of an aspartame beverage mix
  - Two to four glasses of an aspartame hot chocolate
  - Up to two servings of aspartame sweetened puddings and gelatins
  - Several sticks of aspartame chewing gum

- A 34-year-old receptionist was incapacitated by multiple complaints, which promptly subsided after she stopped ingesting such products. Her <u>daily</u> consumption included

- Six 12-ounce cans of aspartame cola beverages
- One 2-liter bottle of aspartame soft drinks
- Five to six packets of an aspartame tabletop sweetener
- Several glasses of presweetened iced tea
- Eight glasses of aspartame hot chocolate
- Eight servings of aspartame puddings and gelatins
- 25-40 sticks of aspartame gum
- 12-24 packs or teaspoons of other aspartame-containing products

- A correspondent working in a large hospital complained to a colleague about her severe headaches, fatigue and memory loss. She was astounded by the friend's first question, "How much aspartame do you use?" The woman wrote me,

> "I always joked that I am a volunteer guinea pig in that I ingest incredibly large amounts of an aspartame sweetener — averaging 20 packets daily — plus large amounts of aspartame drinks, puddings and jello."

There's one figure I have yet to find. It is the number of times the earth would be encircled if all products containing aspartame were aligned. (I welcome the answer from any reader privy to this projection.)

All the Diet Cola Fit to Fly

The late Christina Onassis provided a striking, albeit unfortunate, instance of extreme aspartame consumption. This world-class heiress had her favorite brand of diet cola flown from New York to Paris on her private jet at $300 per bottle (<u>Chicago Tribune</u> November 27, 1989). To ensure freshness, and perhaps as a matter of caution, only 100 bottles were shipped at a time.

Such considerable intake could be logically equated with an "addiction." This analogy is reasonable in light of the severe withdrawal symptoms sometimes encountered when individuals stop using aspartame products (see 10).

Not the Real Thing

The enormous promotional efforts for popular "diet" cola beverages have included contests offering large prizes. They occasionally generated embarrassment.

> Four persons in Michigan (three from Traverse City) thought they won $25,000 Super Bowl prizes in such a contest. Each claimed to have collected all five game pieces from bottle caps and cardboard packages. But corporate officials doubted whether some pieces were "the real thing," particularly one marked "XXII" (The Palm Beach Post December 8, 1990, p. A-10). It seems only five of this genre had been printed.

More on Diet Cola Promotional Duels

In order to maintain a high level of consumerism for their diet colas, two large producers waged costly marketing battles. Such PR events reached fever pitch for the 100 million-plus viewers watching Super Bowl games. Indeed, the gimmicks used in these half-time confrontations sometimes proved more impressive than the game.

> • One featured the first TV commercial with three-dimensional sound technology. Hitch: only 19 percent of U.S. households having stereo television could hear the ads.
> • This competition spawned high-tech games. One used encoded pieces and a scrambled TV signal, along with a special signal enabling viewers to see what they won. The grand prize to "Crack the Code": $1 million!

A peak in oneupmanship occurred when General H. Norman ("Stormin' Norman") Schwarzkopf was shown sipping from a can of Diet Pepsi®. Garry Trudeau described this corporate victory of the Persian Gulf marketing war in the March 21, 1991 edition of the <u>New York Times</u>. The background provides an intriguing story.

> Much of the Arab world expelled the Coca-Cola Company during the late 1960s because it had the audacity to open a bottling plant in Israel. This left the field wide open for PepsiCo, which proceeded to build 24 bottling plants in what would become the future war zone. Under an exclusive military contract, the company then began cranking out 12 million cans a month.

But there's always the other side in Middle East confrontations...including cola wars. It seems that Coca Cola Co. revealed its 106-year-old Coke recipe to several Israeli rabbis in order to gain "kosher" certifications. They were pledged to secrecy, by signing a nondisclosure affidavit (<u>The Wall Street Journal</u> April 29, 1992, p.B-1).

Cartoon 20-2

©*1991 <u>The Palm Beach Post</u>. Reproduced with permission of Mr. Don Wright.*

More Variations On the "Little Lite" Theme

Retailers of many other products learned lessons from the foregoing diet cola pitches. They regarded them as ways of getting one foot in the consumer's door when the economy was "weak."

Cartoonists saw and exploited the inherent possibilities. This is illustrated by Don Wright's rendering in the May 3, 1991 edition of The Palm Beach Post (p. A-10). (Cartoon 20-2)

American Consumer v. American Citizen

Darden Asbury Pyron, Professor of History at Florida International University, noted the various meanings of "to consume" in The Palm Beach Post. They included "to destroy; to spend wastefully; to squander; to engross." He discussed the continuing erosion of our nation's noblest heritages therein on the occasion of the Constitution's 200th anniversary. His comments were triggered by a cinema feature of Mt. Rushmore depicting Thomas Jefferson with a diet cola, and George Washington holding a regular cola. Professor Pyron aptly wrote (reproduced with permission)

> "The title of consumer has almost completely supplanted that of citizen in contemporary American society...However one cuts it or describes it, citizenship is the antithesis of consumption...From its inception in the polis and political order of classical Greece, citizenship has always entailed the subordination of our appetites and individual whims to the law, the larger good, the commonwealth...citizenship is not a salable commodity. The Constitution is not a joke. And Mr. Jefferson is not a diet cola..."

Color-Blind Cola

Aren't cola drinks supposed to be brown...just like roses are red and violets blue? Not necessarily.

The Pepsi-Cola Company announced on November 25, 1991 that it might market a clear, colorless version of its flagship soft drink as a "new age" beverage. While industry analysts studied the prospects, I braced myself for another seeming inevitability: the introduction of an aspartame "light lite" cola.

21

PROMPT FEEDBACK

I predictably hear from listeners and readers after urging caution about the use of aspartame products. Some of their sobering responses would qualify for awards in a competition for "bad news travels fast."

The American College of Legal Medicine

ACLM is not a radio station or an adult congregate living program. These are the initials of the American College of Legal Medicine. Its membership is unique because virtually all Fellows have <u>both</u> M.D. and J.D. degrees.

I was invited to address the 1990 annual meeting of the ACLM, held in Orlando, on the medical and legal ramifications of reactions to aspartame products. Interested members and their wives surrounded me after my presentation.

- One Fellow stated that these products predictably and promptly upset his stomach.
- Another physician had been begging his 35-year-old wife, an R.N., to stop her heavy consumption of diet drinks when she evidenced otherwise-unexplained confusion and memory loss. His pleading remained futile until asking me to confirm this likely relationship in her presence.
- Even the secretary of the ACLM was impressed by the interest I had generated. This young woman then confided <u>her own</u> extreme case of intolerance to aspartame beverages.

A Surprised Niece

Suzie, my wife's beautiful niece, lives in Wilmington, North Carolina. One evening, when there was no answer at our home, she called my mother-in-law <u>very</u> excitedly. The 11 P.M. news on her local ABC affiliate station had just given prominent coverage to my interview during a press conference in Dallas (see 36).

Suzie mentioned her initial reaction: "Golly, that fellow looks familiar! I think I'll call my uncle and see if he knows him." She then realized that the "him" was me!

More Feedback

A long-time patient called my nurse the day after one of my exchanges with Dr. Moser of The NutraSweet Company (see 17). This soft-spoken lady stated, "I'm <u>so</u> proud of Dr. Roberts. But that company doctor wasn't at all nice. Why, he even used some nasty words! I'm glad Dr. Roberts kept his cool."

I received similar responses from persons who were neither patients nor relatives.

> A man who had suffered several reactions wrote that they were incorrectly diagnosed as "transient cerebral ischemic attacks" after the ingestion of aspartame-containing hot chocolate. He continued: "Now I'm OK. But I'm convinced you're right about aspartame. Tell Dr. Moser these are not just 'anecdotes!'"

Feedback to a Spouse

My wife was approached by a friend after I was televised on the Christian Broadcasting Network (see 14). Several persons already told her about seeing me on <u>The 700 Club</u>. This conversation had a unique twist, however, because the friend was a devout Jew.

> "Carol, I must tell you about a remarkable freak coincidence. I was changing channels on my TV set, and happened to see someone who looked familiar. Of all persons, it was your husband being interviewed on CBN, a channel I <u>never</u> watch!"

Extraordinary Feedback

I attended a seminar on diabetes mellitus in Maine during July 1991. The faculty included several distinguished professors.

The case of a young diabetic woman was presented on the second day by her Bangor physician. This recently engaged patient candidly discussed several problems, especially the considerable gain of weight, difficulty in diabetes control, and bulimic behavior. The faculty queried her at length. The moderator then asked, "Any questions from the audience?"

I hesitated because of the predictable response, but then took up the challenge. "How many diet sodas do you drink?" The professors looked at one another on stage and literally wondered aloud, "Why in the world would he ask that question?" The patient promptly replied, "An average of four or five cans a day."

I didn't feel it appropriate to lecture on the adverse influences of aspartame products that were pertinent to her case. Furthermore, I had been interviewed on national ABC television the day before.

We attended a classic Maine clambake that evening. Just before heading for a table near the lovely lake, the executive in charge of arrangements looked at my badge. "Why, you're the Dr. Roberts who made the headline this morning about the hazards of aspartame!"

Carol later casually relayed this exchange to the wife of a physician seated at our table. Her immediate response stunned me. "You're absolutely right! My father is a diabetic. He totally avoids all aspartame products because they totally throw him out of control. We won't even have them in our home!"

I reflected on the extraordinary contrast of these events encountered within a period of several hours. On the one hand, a group of diabetes experts had no inkling of this potentially enormous problem. But the daughter of a diabetic patient was highly attuned to its possible harm for members of her family.

22

CHUCKLES FROM THE THIRD DEGREE

Thomas Moore once observed, "More is missed by not looking than by not knowing." This perceptive adage has served me well for decades.

As persons increasingly kept contacting me about their reactions to aspartame products, I realized the need for compiling this information in some standardized way. Admittedly, it was "anecdotal" — a favorite denigrating term used by the industry and its advocates to indicate case reports not backed up by scientifically controlled studies. I therefore drafted the 9-page questionnaire mentioned in the introductory Overview.

The response to the initial mailing was gratifying. In fact, the "returns" would make most political pollsters livid with envy. About 31 percent(!) of over 1,100 persons in the United States and Canada completed and returned this questionnaire. The batting average for subsequent correspondents has been even more impressive.

Many rib-tickling "pearls" were nestled among the clinical insights so provided by this heterogeneous group. I will share a few.

"General Information": Age and Occupation

Mundane items dealing with age and occupation occasionally evoked a smile.

- A 92-year-old woman had remained remarkably active...even while confined to a wheelchair after her recent leg fracture. The ingestion of an aspartame product predictably precipitated severe vomiting and diarrhea. These complaints stopped when she resumed sugar. Wishing to emphasize that her brain was not senile, she wrote: "I _am_ in my right mind."
- A woman from Parkers Prairie (Minnesota) described her occupation as that of a "Junque and

Anteek Dealer."

• Debbie assumed the role of "aspartame victim" at a press conference sponsored by the Aspartame Consumer Safety Network in Dallas (see 42). This young woman described her previous severe grand mal convulsions when using aspartame products. Her business card read: "Total Hang Ups: Installation of All Window Treatments."

Completed questionnaires were received from many professionals. They included doctors, dentists, nurses, lawyers, school administrators, social workers, college professors and even priests. But the one relayed by a 29-year-old field director of the Girl Scouts made me feel that my efforts had reached mainstream America.

The Matter of "Gender"

Females outnumber males by a 3-to-1 ratio in virtually every category of adverse reactions to aspartame products. I have elaborated on this remarkable phenomenon and its likely explanations in previous publications.

This is one area wherein I tread <u>very</u> carefully when discussing data. Having a wife who is a politician (see 6), I consciously head off any criticism about male chauvinism by using the term "gender" instead of "sex."

Misspellings

My ventures into the epidemiology of aspartame reactions reinforced the great need for national emphasis on spelling. The following are examples of misspellings. However, they did not diminish my gratitude or respect for these cooperative respondents.

• "I had an unexpected sizure."
• "I was looseing weight."
• "I spoke to a drugest about it and he did'nt beleave me." (Her husband also developed "dyseentary" from aspartame.)
• "I started haveing cramps and dysentary after using

118

aspartame. After hearing it caused dysentary and epilesy in some people, I stopped useing it. The very next day, I did not have any cramps."
- "My daughter's bigest problems was dizzeness."
- "I am a 35 year old deviorced and educated women."
- I've been treated for heart palpulations and blured eyesight."
- "Dieing"; "side affects"; "labratory"; etc.

One must exercise caution in discussing this subject. Others, especially politicians, have discovered the inherent hazards of doing so.

President Theodore Roosevelt attempted to simplify words by eliminating silent vowels. He even ordered the public printer (but while Congress was not in session) to change the spellings of "kissed" to kist," "blushed" to blusht," "gypsy" to "gipsy," "through" to "thru," and whiskey" to "whisky." Other innovations by T. R. were "surprized," "compromize," and "artizan." (They also have been advocated for 84 years by Th Simpliyd Spelng Society in an attempt to purge many redundant letters from English words that "are linked to economic and social problems in Britaina nd America.")

The ensuing clamor for "orthography," as prescribed in conventional dictionaries, was successful when journalists decided to return such fire with their literary fire. The editor of the Louisville Courier - Journal wrote

"Nuthing escapes Mr. Rucevelt. No subject is tu hi fr him to takl, nor tu lo for him tu notis. He makes tretis without the consent of the Senit. He inforces such laws as meet his approval, and fales to se those that du not soot him. He now assales the English langgwidg, constitutes himself a sort of French Academy, and will reform the spelling in a way tu soot himself."

Labeling

Scores of correspondents attacked the confusing labeling and "fine print" found on some aspartame products. These problems are similar to those involving other ingredients and additives.

Cartoon 22-1

The national outrage about confusing labels, fostered by the gospel of deregulation, peaked over the issues of "low fat," "no cholesterol," and "high fiber." Unable to cope with the FDA, the regulators of nine states formed a task force, duly captioned "food police," and sued manufacturers for misleading labels and ads. (Time July 15, 1991, p. 58)

One person bemoaned the practice by some local merchants of putting price tags on the caps of soda bottles where the ingredients were listed. She added, "Furthermore, they are written so small you need a magnifying glass to read them." Others echoed this sentiment for a variety of additional aspartame products.

The Matter of Logos

My enjoyment of good cartoons (see 3 and 4) is shared by many aspartame reactors. Several even submitted their own creations in letters as part of the completed questionnaire. These artistic efforts apparently enabled them to vent longstanding frustration and anger. One hit the bull's eye, literally, with his alternative "swirl" suggested as a "NO" logo to warn consumers.

As awareness of severe reactions to aspartame products increases, PR agencies might find themselves back at the drawing board to devise acceptable substitutes for the "swirl." Lest this seem farfetched, Dear Reader, just recall the dilemma of "Old Joe" ads for Camel cigarettes.

On March 9, 1992, U.S. Surgeon General Antonio Novello supported the American Medical Association in urging the R.J. Reynolds Company to halt such promotion. Why? Studies revealed that children were more familiar with this cartoon camel than with Mickey Mouse or Cheerios! Furthermore, Camels sales to adolescents had surged after this character was introduced in 1988. If poor Old Joe were to hit the dust, would the "swirl" appearing on thousands of popular "lite" products also be on the endangered list?

Logo logic — or more accurately nonlogic — involving other products and firms has made the headlines.

A bearded man-in-the-moon logo had been the emblem for <u>Procter & Gamble</u> since 1851. Its products came under heavy attack when this trademark was interpreted in some quarters as a satanic symbol. Satan-rumormongers then had a field day when told to press "6" after calling the company. This apparently provided them with the ultimate proof needed for true believers of the Book of Revelation, which linked "666" to the Devil. P&G threw in the towel. Its redesigned logo eliminated curls from the beard of the man in the moon.

"Comments"

The questionnaires completed by reactors to aspartame products contained scores of interesting critical "comments" and "bones of contention." Here are a few.

- A 39-year-old sales representative with multiple severe reactions wrote
  "My body was in <u>excellent</u> shape. That's what hurt more than anything. I had great legs — now I look like <u>hell</u> and am out of tone...and I'm really <u>pissed</u>!! This isn't fair."
- A 32-year-old chap with grand mal convulsions attributed to using aspartame products also had suffered mood changes. The result: "My wife said she felt alienated from my affections."
- The wife of another fellow described how this "very outgoing and vibrant-on-action man" became trans-

formed into an "in-spirit-a-nothing" person. Furthermore, he suffered attacks of dizziness shortly after ingesting an aspartame beverage during several retrials.

- One woman concluded that drinking diet colas had made her "allergic to the 21st Century."
- A 32-year-old housewife experienced severe irritability and anxiety attacks after drinking diet colas. She stated, "When you take aspartame, you are taking 'speed.'"

Doodling

Some reactors emphasized their "righteous indignation" through personalized doodling. For example, a 33-year-old teacher developed violent headaches after swallowing aspartame products. His response to one of the questions under "OTHER USEFUL INFORMATION," appears in Cartoon 22-2.

Do you **still** use products containing Aspartame?   (Circle) Ye(No)

Cartoon 22-2

Physician Encounters

The greatest frustration of many aspartame "victims" was the encounter of combined ignorance, apathy and hostility to this subject by physicians they consulted. Many singled out neurologists in this respect.

- A woman with multiple incapacitating problems experienced complete relief within ten days after avoiding aspartame products. She actually had been scolded by her physician for stopping his prescribed medications ...even though she continued to remain symptom-free!
- A 31-year old registered nurse experienced her first grand mal convulsion within hours after drinking two liters of an aspartame-sweetened drink. All the appropriate studies proved "negative." Deciding to research this subject in the hospital's medical library while still an in-patient, she found three pertinent articles. When she presented these fruits of her labors to the neurologist, he advised, "Stop reading!"
- Larry Wilson (see 36), 35-year-old nurse anesthetist, developed headaches, memory impairment, visual problems, and three subsequent convulsions while drinking 4-6 diet colas daily. He told a Senate hearing about his neurologist's reply to the suggestion that aspartame might be contributory: "Wouldn't it be a shame if all that is wrong with you is NutraSweet?"
- A 61-year-old forester repeatedly developed grand mal convulsions after drinking aspartame beverages. He suffered a painful dislocated shoulder during one attack after being unknowingly served an aspartame drink. His wife wrote: "I asked the doctor if this didn't prove that it was the aspartame causing his convulsions. He said, 'Maybe.' I said, 'Shouldn't someone write the company and tell them what had happened?' His reply: 'Yeah, you do it.'"
- A 39-year-old homemaker experienced irritability, memory loss, insomnia, and a marked change in personality after consuming aspartame products. She stated, "I was amazed by the total personality change. My tests were primarily my own. Doctors think I'm 'nuts' to think a sweetener would be the problem!"

Precise Tracking

Some aspartame reactors could recall the <u>exact</u> onset and duration of their misery, even before identifying aspartame products as the cause.

A 36-year-old farmer became nearly incapacitated after drinking 8-10 cans of an aspartame diet cola daily for three years. He had consulted more than five physicians for severe impairment of vision, headaches, dizziness, tremors, depression, personality changes, attacks of shortness of breath and palpitations, abdominal pain, and severe itching. He chanced to hear my discussion on CBN (see 14), and returned a completed questionnaire. Under suggestions about the regulation of aspartame products, he remarked, "Should put pressure to remove it from the market. I would rather risk saccharin than go through the past 2 years, 7 months and 16 days again."

Aspartame Activists

A few aspartame "victims" opted to take matters into their own hands.

- A 42-year-old woman predictably suffered severe reactions after consuming aspartame. They consisted of dizziness, confusion, slurred speech, and a rash. She retaliated by wearing a T shirt with bold letters at an ice-skating championship sponsored by the maker of an aspartame product. It read "ASPARTAME IS DANGEROUS TO YOUR HEALTH." She also passed out flyers warning about aspartame.
- Mary Stoddard, an aspartame reactor and President of the Aspartame Consumer Safety Network (see 36), conducted a one-woman educational campaign. Learning that a two-hour class, "Sugar-Free Cooking with Equal/NutraSweet," was scheduled at Brookhaven Community College, she sprang into action. Although 14 persons already had registered,

this class was cancelled after Mary protested to the Dallas County Community College District's Board of Trustees (The Dallas Morning News March 28, 1989, p. A-16).

Suggestions about Regulation

My questionnaire ended with the request that correspondents offer ideas "about regulating the use of aspartame." Some were mentioned earlier. Here are a few more blunt replies.

- "Tell the truth!"
- "This product may be hazardous to some people."
- "What we need is another Boston Tea Party!" (This suggestion came from a school teacher who had severe reactions to aspartame-sweetened tea).

Attitudes Of, and About, the FDA

This subject was a real bone of contention by aspartame reactors. It will be discussed in Chapter 42.

A New Census?

My first interview about adverse reactions to aspartame products was shown on West Palm Beach TV Channel 5.

The office manager of a large local firm called the next day, and requested 90 copies of the questionnaire! While my staff ordinarily handles such requests, I followed through personally on this one because it begged for clarification.

The manager explained that she was in charge of machines dispensing soft drinks. In this capacity, she had observed striking changes among certain employees who were consuming considerable amounts of diet colas. They included erratic behavior, depression and other severe emotional problems. But her real interest in the matter boiled down to improving efficiency and productivity rather than altruistically pursuing an epidemiologic research study.

"Dept. V"

An Illinois optometrist became enthralled by my observations concerning ocular problems encountered in aspartame reactors... especially dry eyes and irritation from contact lens (his specialty). He carefully studied my writings on this subject.

With permission, he condensed my questionnaire into a one-page survey intended for interested eye-care professionals and professional journals. In addition to visual complaints, it contained other symptoms frequently experienced by aspartame reactors.

The bottom of this shortened version read

"Thank you. Please send completed form to
H. J. Roberts, M.D.
Palm Beach Institute for Medical Research
Aspartame Study (Dept. V)
300 27th St.
West Palm Beach, FL 33407"

This optometrist anticipated my question about "Dept. V." He indicated that the "V" meant "Vision." He magnanimously added, "I'll let the ear people have 'E.'"

23

ENCOUNTERS WITH THE MEDIA

Let me regale you with some encounters as "a reporter on reporters" — or as "a medium for the media." They include an assortment of newspaper reporters, feature writers, radio and television talk show hosts, PR firms representing corporate interests, and even a slide projectionist.

An introductory caveat is appropriate: granting interviews with media persons could be risky business when they perceive themselves as "the message" rather than messenger. The possibility always exists of being misinterpreted, misquoted, or making some stupid remark while being badgered.

Media Indoctrination

I share certain attitudes with most physicians about upholding the dignity of our profession. They concern honoring its traditional lofty principles, and avoiding self-serving publicity.

> The Lancet, a prestigious journal, discussed "medical men and the lay press" in its April 24, 1915 issue (reproduced with permission).

> "We are at all times of the opinion that medical men writing upon medical subjects in the lay press should exercise great care and discretion in their choice of subjects and in their manner of approaching them. Advice given in popular health which will tend to the preservation of the public health can be very useful, and information as to the nature of a particular disease if it will have the effect of sending patients in the early stages to seek medical advice — general medical advice that is, and not advice from Dr. A or Mr. B — may not be amiss in some instances."

My motives were challenged by industry and industry-supported organizations after becoming interested in adverse reactions to aspartame products. If I had to identify some milestone event as such a "consumer advocate," it would be the press conference on July 30, 1986 called by Aspartame Victims and Their Friends. Facing a barrage of TV cameras and newspaper reporters, I summarized the disturbing data on my first 100 aspartame reactors.

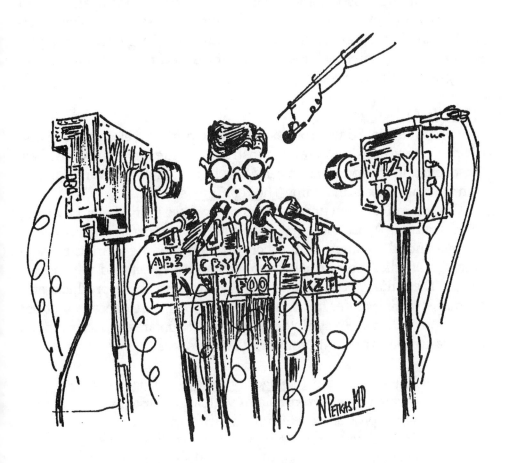

Cartoon 23-1

Subsequent events took a remarkable twist. Most investigators in academia who had written on this subject chose to sequester themselves from the media. As a result, I found myself virtually alone in ongoing altruistic confrontations with well-rehearsed and well-paid representatives of a billion-dollar industry.

The sense of being a "majority of one" crystallized on October 16, 1986. The occasion was a national press conference held at the Dirksen Senate Building. Aspartame Victims and Their Friends again invited me (at my own expense) to present my updated researches. This meeting became my first encounter with several "front" organizations representing "vested interests" (see 15) that preached both the virtues and safety of aspartame products.

A Media Affliction

Let me be fair. Some professionals in the print and electronics media expressed extraordinary and constructive interest in my studies. Unsolicited invitations were received from prominent radio and television talk show hosts, including The 700 Club (see 13).

One reason for such unusual interest by these individuals, however, soon surfaced. Most had personally suffered confusion, memory loss, serious mood swings, and other problems while drinking large amounts of "diet" soft drinks, or coffee and tea sweetened with aspartame. Their thirst became intensified both from these products and profuse sweating under hot lights. Several hosts were concerned about the serious threat these symptoms posed to their careers. Accordingly, these case reports are humorous only in retrospect.

- The 40-year-old manager of a news service became aware of his own problems with aspartame products while producing a television program on this subject! He had been consuming two large (two liter) bottles, and up to ten regular glasses of diet sodas daily. His complaints included severe dizziness, intense thirst ("the worst of all my complaints"), recent "dry eyes," difficulty wearing contact lens, insomnia, nausea, diarrhea, severe itching, frequent urination at night, "low blood sugar attacks," and pins and needles in the arms and legs.

- A 31-year-old <u>radio talk show host</u> had been drinking three cans of diet colas daily in an attempt to avoid gaining weight. He experienced "periodic, uncontrolled outbursts of anger at work, depression, and suicidal thoughts." He also noted "loss of control over moods," and the feeling of being "pulled down." In attempting to remain professionally objective, he stated

  "It's hard to attribute my reactions solely to the use of aspartame. However, the most severe reactions appeared as I began its use, and lessened considerably or disappeared after this use was stopped."

- A 27-year-old female <u>television producer</u> consumed three cans of aspartame-containing soft drinks and one glass of presweetened iced tea daily over a period of two years. She developed pain in both eyes, severe headaches, tingling of the extremities, palpitations, nausea, and marked frequency of urination. A CT scan of the brain and various eye tests proved normal. She deduced a clearcut association of these complaints with aspartame products when her health improved after avoiding them.

- A 30-year-old <u>employee for TWO radio stations</u> suffered severe problems involving memory, confusion, depression (including suicidal thoughts), a change in personality, insomnia and dizziness. She consumed one packet of an aspartame tabletop sweetener in each cup of coffee. Her symptoms improved within three days after she stopped using the sweetener, and did not recur. She refused to retest herself. She expressed her indignation in these terms:

  "I become angry when I see commercials promoting aspartame. How harmless they make it all look. The all-American brown & white can and something about bananas. If this product is truly harmless, why the marketing aspect?"

Media Poetic Justice

Aspartame reactors who were members of the media vented their frustration over the indifference to this matter by colleagues and listeners.

- A TV station employee experienced repeated severe headaches, nausea and fatigue from aspartame products, as did her sisters. She was outraged over "the medical reporter's refusal to recognize aspartame reactions as widespread and serious."
- Neil Rogers, a South Florida talk show host (see 7), told skeptical listeners about his multiple reactions to aspartame products. He then offered this well-known advice: "If in doubt, throw it out!" (The USDA issued this same warning for homemakers who suspect the safety of canned foods.)

Talk Show Callers

I accepted invitations from radio talk show hosts as far away as Toronto for good reason...the extraordinary input provided by callers who suffered aspartame reactions. Indeed, several of these conversations provided my first inkling of a previously unsuspected clinical problem.

Some of these discussions were poignant, others unexpectedly dramatic. One occurred on September 3, 1987 during an interview on WJNO, a West Palm Beach radio station.

> The caller stated that he was confined to bed for several hours following a lumbar puncture that morning. It was part of an ongoing evaluation for his unexplained impairment of vision.

> This chap became a heavy consumer of aspartame drinks after moving to Florida the previous year. He experienced significant deterioration of his sight thereafter. He consulted several ophthalmologists locally and at an eye center in

Miami. Numerous studies — including CT scan and MRI studies of the brain, an electroencephalogram, and multiple eye tests — were normal.

A neurologist then recommended a lumbar puncture "for the sake of completeness." Advised to remain in bed after this procedure, the patient chanced upon WJNO. He expressed astonishment on hearing about aspartame-related eye problems since no physician had ever mentioned this possibility!

I refused to let such talk-show "popularity" interfere with my obligations as a primary care physician. The orbiting of other doctors on this tangent captured the eye of cartoonists.

One focused on a famous "diet doctor." His receptionist told a prospective patient that he was booked for the next three months ...on talk shows.

700 Club Fame

I came to realize the national impact of the Christian Broadcasting Network — first after being televised on The 700 Club, and later during an interview on its sister radio program (see 14). The special interest of CBN in this realm reflected Reverend Pat Robertson's unequivocal association of his severe confusion and memory loss with the previous drinking of aspartame sodas. The number of my patients and other acquaintances who saw or heard me on CBN shows (see 21) proved mindboggling.

These appearances did not go unnoticed in other quarters. The sergeant-at-arms of my Rotary Club regarded them as golden opportunities to levy a stiff fine for alleged "promotion."

The Enthralled Projectionist

I gave my first formal address about this subject to the Section on Internal Medicine of the Southern Medical Association in Atlanta on November 10, 1986. It entailed a review of 360 aspartame reactors in my data base at the time.

The large convention room for this session was packed with physicians and members of the press. During the question period, The NutraSweet Company's representative (see 17) embarked on a lengthy "rebuttal"...to the dismay of a timid chairman. I challenged each of his presumed "facts" in the ensuing debate.

> Let me explain my seemingly inappropriate smile during subsequent comparable discussions. It was triggered by recalling this little ditty concerning corporate party lines: "When there's debate, obfuscate!"

Another memorable event occurred two hours later. A fellow approached me when I arrived early for the afternoon program. He asked, "Weren't you the doctor who talked about NutraSweet this morning?" I replied in the affirmative. The person then introduced himself as the slide projectionist, and made a comment I hadn't heard from a projectionist in nearly four decades of giving lectures: "That sure was an interesting lecture, Doc!" He added, "And from now on, I'm going to be real careful about those diet drinks!"

The Case of A "Nervous Nellie"

One of South Florida's public television stations requested an interview following publication of my first book on adverse reactions to aspartame products. I sandwiched it in on a Wednesday — jokingly called my "day off." The producer said that this program would be shown "sometime in the near future."

My secretary received a call one month later from someone at the station. The message: my interview was to be featured the following Saturday at 6:30 P.M.

The program host called Saturday morning. Hesitatingly, she informed me that a "Nervous Nellie" in the higher echelons was concerned about the station's legal vulnerability in the event sales of aspartame products declined in South Florida. I replied

> "Just tell your Nervous Nellie that I have a constitutional right to express my opinion as a concerned

physician, an informed scientist, and a citizen. You also might tell her that there is as much a chance of litigation as my suing your station for watching the lips of President Bush when he asked the nation to read his lips about 'no new taxes'...and now finding that we will be paying more in taxes because he changed his mind."

Another message followed within the hour. "Everything's fine! Your interview will be aired as scheduled." There was one hitch, however. Nervous Nellie offered the manufacturer "Equal® time" on a subsequent program.

No Strange Bedfellow

The cliché that politics make strange bedfellows hardly needs any explanation. This term unexpectedly surfaced during an interview about aspartame with a newspaper reporter. The alleged political "bedfellow" in this instance was none other than my wife (see 6)!

I happened to use the term "corporate neutral," meaning I had not received financial support from companies having vested interests (see 15) for my studies on aspartame reactions. The reporter then remarked: "But Mrs. Roberts got a $500 campaign contribution from the sugar industry in 1986." I promptly replied to this hostile query, which clearly had been researched. "Young lady, if you think that Mrs. Roberts' vote can be brought for $500, or even for $500,000, you <u>certainly</u> don't know her!"

A Familiar Name

Ellen McGarrahan is a reporter for <u>The Miami Herald</u> who once covered Palm Beach County. She requested an interview after the publication of my first book on aspartame reactions.

Ellen's two-page feature appeared in the January 5, 1990 edition (pages E-1 and E-5) under the caption, "A Sour View of a Sweet Subject." The many details — based on additional interviews with The NutraSweet Company, the Community Nutrition Institute, and the FDA — underscored her professionalism.

Four months later, a headline in the May 1, 1990 edition of <u>The Mi-</u>

ami Herald caught my attention. It read: "Governor Dismayed By Press Aide's 'Joke.'" The article noted our Gov's dismay about a staff member "who jokingly asked a newspaper reporter who she slept with to get her job."

Intrigued, I read on. The reporter was Ellen McGarrahan, who had been transferred in the interim to the paper's bureau in Tallahassee. (Considering all the inconvenience and hassles involved in this two-month beat to cover the annual session of Florida's Legislature, my politician-wife didn't regard such an assignment as a significant promotion.)

Ellen retaliated with literary fire. Reporting the foregoing comment in Talk of Tallahassee, she emphasized that men still dominate politics in our State's Capital. Acknowledging that the press aide had been joking, she nevertheless countered that his male-boss probably wouldn't appreciate a similar accusation.

Barbara Mullarkey (see Figure 27-1)

The writings of Barbara Mullarkey came to my attention shortly after an interest in aspartame reactions began to evolve. She was a home economist, author of a biweekly column in the Wednesday Journal dealing with nutrition and ecology, nutrition editor of Conscious Choice Magazine, and editor/publisher of NutriVoice (a health watch newsletter).

Over the ensuing years, my admiration for Barbara's determination and courage as a "truth-seeker" increased. She probed and probed...refusing to be put down by semantics, especially the "reporter" label. Barbara once retorted, "There's a big difference between a reporter and a columnist. I'm a columnist!"

Not surprisingly, Barbara drew the ire of some corporate executives. One answered her explanation about "seeking the truth" with the unflattering remark, "Bull..." She responded, "Now is that any way to treat a Mullarkey mother of five?" (Wednesday Journal, March 5, 1985, p.19).

Barbara became a preferred target for criticism and ridicule by Dr. Robert M. Moser, Vice-President of The NutraSweet Company (see 17). He commented on her column describing the November 1991 symposium on aspartame safety at the University of North Texas (see 11):

"She (the coordinator) did assemble enough for eight hours of harangue; it was replete with the usual barrage of anecdote, innuendo and opinion familiar to those who read Mullarkey. Not one shred of scientific evidence was presented...The column by Mullarkey displayed unusual — even for her — contempt for fair play. She violated a basic tenet of good journalism; she quoted me out of context — in an effort to make me appear duplicitous" (Wednesday Journal December 11, 1991).

What was the breach of good journalism by this pro? Barbara had restated Moser's own words from The Pharos (Summer 1991), journal of Alpha Omega Alpha (an honor medical society), based on his talk to Yale medical students.

"If you *hate* people, you can spend your life plugged into a mountain of shining diagnostic equipment or perched beside an electron microscope. Or you can become a cruise-ship doctor and play patty-cake with rich, blue-haired widows, or you can become a department chairman at Yale, or you can work for The NutraSweet Company."

Lest I also be accused of quoting my former colleague out of context, the preceding paragraph read

"If you *love* people, you can put your all into private practice, or work in an AIDS clinic, or labor on behalf of the World Health Organization in Sudan, Ethiopia, or Southeast Asia, and help sick, starving, frightened people."

Barbara remains suspicious of professionals and bureaucrats who always defended aspartame. Her suspicions became compounded if the same people also consumed significant amounts of aspartame products. One was a new FDA commissioner who usually started his day by drinking a diet cola.

The refusal of companies to give the chemical source(s) of aspartame became another pet peeve for Barbara. As far as she was concerned, this had the validity of saying that milk comes from a carton.

Reference to aspartame as a "natural" product proved one more waving red flag for Barbara. She commented, "Arsenic also is natural."

This "tough" columnist developed enormous compassion for aspartame reactors who became the object of corporate scorn and jokes after deciding they weren't "going to take it (disinformation) any more." One was Shannon Roth, whom a company attorney sarcastically called "Blinky" (Wednesday Journal November 20, 1991). This young woman founded Aspartame Victims and Their Friends after being blinded in one eye following the ingestion of considerable amounts of aspartame products.

Barbara leveled her sharpest professional criticism, however, at other columnists who remained true believers in the matter of aspartame safety. One nationally-syndicated feature writer wrote her, "My support (for these products) is firm and unequivocal."

Barbara's dialogues and columns were sprinkled with great expressions. One made a lasting impression on me. "If you know the truth and don't speak it, you're a liar."

Discussing Real World Research on *Real World*

I have emphasized, and reemphasized, a key objection to reports by researchers proclaiming their inability to reproduce serious reactions allegedly caused by aspartame products. It focuses on pure aspartame administered as capsules or as freshly prepared cool drinks to such persons. What's my hangup? These ploys conveniently sidestep the intake of several potentially harmful breakdown products of aspartame resulting from prolonged storage and exposure to heat.

I also repeatedly used the term "failure to give real-world aspartame products," or a slight variation thereof, when discussing this matter in my writings and lectures on aspartame disease.

I received a call from a radio producer on Thursday, April 2, 1992. He wanted to interview me the next day for his national weekly radio program, *Real World*. It would be then broadcast on Saturday to 280 stations via satellite. I agreed.

During my interview, the co-anchor asked about flawed research involving this "most tested additive in history." I mentioned the failure to test "real world" products taken from market shelves in such presumed rechallenge studies. A humorous thought then took over. "Your program seems to be the logical forum to raise this issue for one very good reason, because it's called *Real World*."

Some Basic Questions

As my studies and publications on aspartame disease expanded, I found myself increasingly challenged by variations of two queries from media interviewers.

- I. "Dr. Roberts, you have spent an enormous amount of time, energy, and even your own money to pursue this subject. But you obviously have other important commitments to an active medical practice and your family. WHY??"

- II. "Who should the public believe — professors and researchers at major universities or institutions who state unequivocally that aspartame is safe, or Dr. Roberts?"

I indicated that both questions were proper and deserved honest answers. The latter actually had been forged from longstanding self-analysis concerning these issues. My replies to the second question generally combined the following attitudes:

- "I have always attempted to convey my findings and opinions as accurately as possible over four decades of involvement in clinical research. My conscience would not permit me to do otherwise, especially because many persons trust my work."

- "I will gladly modify, or even retract, any statement I have made if a colleague who has worn the same moccasins can show me where my patients' experi-

ences or my own observations were wrong. Furthermore, I shall extend the real appreciation a teacher
deserves."

The first question, however, proved the troublesome in the context
of yet another question: "Has the medical and scientific community
reached the point where research is suspect because it was done out of
altruism, and supported by the investigator's own resources rather than
a corporate or governmental grant?"

I usually paused and limited my response to a smile...anticipating
that I might end up giving a sermon. Let me now impart several of
these unpreached messages to any frustrated interviewers.

- "If not now, when?"
- "Sometimes one must rebel to excel."
- "A man with commitment can become a majority."

24

ON THUNDEROUS THIGHS AND LUSCIOUS EARLOBES

American women are consuming artificially-sweetened foods and beverages in "prodigious" amounts (see 20). To a large degree, this reflects an obsession of epidemic proportions that is aptly designated "fear of fat." Instead of stoically acknowledging that members of their gender come in all shapes and sizes, many females opt for the "thin and perfect" prototype of slim actresses and models (the "Twiggy look.") Some justify the associated hunger and emaciation through the perception of gaining power by such self-discipline. Conversely, full-figure contours reminiscent of the Gibson Girls are despised.

The inimitable Ogden Nash opined on this matter in <u>Just a Piece of Lettuce and Some Lemon Juice, Thank You</u>.

> "Though human flesh can be controlled,
> We're told, by this and that,
> You cannot win:
> The thin stay thin,
> The fat continue fat."*

Syndicated feature writers have been able to pay the rent by focusing on this unrealistic pursuit of sylphlike thinness.

> • Erma Bombeck referred to women who worship "at the shrine of perpetual lettuce." Her attitude was influenced by contemplating the refusal of figure-conscious women on the Titanic to indulge in any desserts.

---

* From <u>Face Is Familiar</u>. © 1933 Ogden Nash. Reproduced with permission of Little, Brown and Company.

- Ellen Goodman pointed her talented finger to an associated contemporary affliction: "the five-pound overweight neurosis."

Ogden Nash had similar thoughts about these famous five pounds in <u>Curl Up and Diet</u>.

"Some ladies smoke too much and some ladies drink
    too much and some ladies pray too much,
But all ladies think that they weigh too much.
They may be as slender as a sylph or a dryad,
But just let them get on the scales and they embark on a
    doleful jeremiad;
No matter how low the figure the needle happens to
    touch,
They always claim it is at least five pounds too much;
No matter how underfed to you a lady's anatomy
    seemeth,
She describes herself as Leviathan or Behemoth;
To the world she may appear slinky and feline,
But she inspects herself in the mirror and cries, Oh I
    look like a sea lion."*

More Fodder For Cartoonists

Observant cartoonists also exploited this pervasive dilemma in our society.

- Lampooning the theme, "A mind is a terrible thing to waste," one strip focused on "The waist is a terrible thing to mind."
- "Throw away" medical journals repeatedly feature cartoons on numerous variations of these themes. Example: as two women are indulging in dessert, one says, "Going on a diet always seems to give me an appetite."

---

* From <u>Verses From 1929</u>. ©1932, 1935 Ogden Nash. First appeared in <u>The New Yorker</u>. Reproduced with permission of Little, Brown and Company

- Cartoons in popular magazines focus upon the admonition of doctors by irate plump female patients. Example: "You mean to say that with all the progress in medicine I keep reading about, the only advice you're able to give me is to eat less?"
- The assertion that "fake fat" (see 43) can cut down calories by 80 percent or more provided imaginative cartoonists with additional manna. Example: a corpulent food addict rejoices before the boob tube with the new excuse, "Now I can eat 80 percent more!"

Rebellion In the Ranks

The agony resulting from such personal and promotional deception can be monumental. After years of dieting, one teenager regarded the quest for thinness as unattainable Utopia. Life had become a literal hell as she endured the "punishments" of loneliness, pity, self-hatred and loathing because of her belief that our society doesn't embrace persons who are slightly plump.

Some disenchanted "born fat" individuals don't pull their punches in criticizing the diet industry as "a fraud." Individual who lost "tons of fat" on all kinds of diets and other professed "metabolic magic" are in the forefront. Perhaps it was someone in this category who punned, "What foods these morsels be!"

As far as persons in this category are concerned, there is little difference between a "fat chance" and a "slim chance" for success...notwithstanding the impressive before-and-after photos displayed in ads. Indeed, such promotional tactics serve as reminders of the old saw, "There's no such thing as a free lunch."

> Suzie Heyman, author of Diet Is a 4 Letter Word (Starlight Publishing), asserted: "They make millions of dollars on the desperation of fat people with programs and diets that do not work" (Palm Beach Daily News July 14, 1990, p.7).

A 21-year-old University of Connecticut junior received national attention when she refused to turn the other cheek in her version of "the most unkindest cut of all" (Shakespeare, Julius Caesar III:ii). She filed a sex discrimination complaint with the Connecticut Commission on Human Rights and Opportunities. This woman sought reinstatement to the cheerleader squad after being told she wasn't thin enough in spite of virtually starving herself. Her measurements were 5-foot-6 and 130 pounds...just five pounds over the arbitrary weight limit. The latter figure presumably reflected a "risk management" policy aimed at preventing injuries during acrobatic routines. This student made a convincing point for discrimination: the absence of such a limit for male cheerleaders (The Miami Herald August 15, 1991, p. A-5).

Flab Phobia: A Love-Hate Phenomenon

Weight-conscious American women constitute the majority of aspartame consumers. (They also predominate among those having adverse reactions to products containing this chemical, as noted in the Overview.) This demographic fact can be explained by an attitude that is epidemic among females of all ages. I have labeled it the "Think Thin Obsession" (TTO). It even haunts women of average weight who have been endowed with intelligence, and possess considerable self-esteem for what is beneath their scalps.

Damon Hertig captured this theme on the front cover of the November/December 1990 edition of Diabetes Spectrum. It is reproduced in Cartoon 24-1 with permission from the American Diabetes Association.

Cartoon 24-1

© *1990 American Diabetes Association. Reproduced with permission of the American Diabetes Association, Inc.*

Let me be blunt. Physicians, dietitians, sociologists and psychologists cannot come to successful grips with patients who have "obesity" or other eating disorders, 90 percent of whom are female, if they fail to consider the TTO in their deliberations. In effect, American women harbor an extraordinary love-hate relationship about food that far surpasses similar attitudes among European counterparts.

- My daughter Pamela came to appreciate this difference while attending the London School of Economics for an advanced degree. She was amazed at the contrast in attitudes by new friends there and those in the States. For example, conversations with the latter tended to be dominated by the subjects of food and weight.
- Julia Child, our TV doyen of French cuisine, commented: "I think women are acting hysterical about food. You'd think that modern, liberated, educated women would not act like ninnies, but they do" (The Palm Beach Post June 14, 1991, p. D-1). (Also see 7)
- Henry Jaglom drove this point home in the film Eating, "a very serious comedy about women and food." There wasn't one man in the cast!

The problem becomes compounded by the dismal and demoralizing long-term results of dieting, alluded to earlier. Up to 95 percent of dieters regain their weight within one year. I confirmed this truism in the 1950s as one of the original investigators of Metrecal® — or "metered calories."

Many seem to equate aspartame products with the Holy Grail. Unfortunately, this conviction often assumes the attributes of a metabolic-psychologic sting operation when formerly obese women find dietary control difficult without their use. Some even have been willing to tolerate severe headache and other significant side effects associated with such consumption as the price for weight loss.

The Epidemiology of Fear of Fat

Recent surveys reflect the unmistakable extent of the TTO phenomenon.

- Family Circle Magazine found that 78 percent of

American women think they're too fat. This is three times the number who <u>really</u> are overweight according to the norms generally accepted by health professionals.

- More than half of 160 college coeds surveyed by a University of Southern California researcher resorted to unhealthful eating habits in an attempt to become or stay slim (<u>The Palm Beach Post</u> March 19, 1990, p. E-1). In fact, one out of four admitted to engaging in bulimic behavior...including induced vomiting.

- A University of Florida psychologist reported that up to 60 percent of college women have some kind of eating disorder attributable to an obsession with thinness. In her group's study of 1,335 females students at this university, 80 percent stated they had dieted during the past year — even though the average respondent was 5 feet, 5 inches tall and weighed 122 pounds (within the low-to-normal range) (<u>The Palm Beach Post</u> December 1, 1990, p. A-3). Furthermore, five percent described bulimic behavior (i.e., purging at least twice a week), while four percent starved themselves and became anorexic.

- Researchers at the Centers for Disease Control decided to evaluate the eating behavior of 364 female freshman attending "an upper-middle-class private college in the South" with the "desire for a slender physique" (<u>Southern Medical Journal</u> 1991; 84:457-460). These were their findings:

  * Nearly 60 percent were currently on a diet to lose or to maintain weight.
  * 29 percent had used crash dieting or fasting since entering college seven months earlier.
  * 20 percent had used diet pills at some time.
  * 13 percent had resorted to vomiting, or had used laxatives or diuretics at some time.

Sad to relate, the fear of fat is generally entrenched by the time most girls — and many boys — attend <u>elementary</u> school.

- This attitude was found among all age groups in nearly 500 girls attending grades 4 through 12 at San Francisco parochial schools, according to a University of California study (News/Sun-Sentinel of Fort Lauderdale, November 1, 1986, p. F-1). Thirty-one percent of the nine-year-olds actively worried about being overweight or becoming fat!
- Another study of 168 girls and 150 boys attending grades 3 through 6 at two randomly-selected schools in Cincinnati, aged 7-13 years, was reported in the September 1989 issue of Pediatrics (Volume 84, pages 482-489.) (The mean family income was over $41,000; 92 percent were white.) Both genders were concerned about body image even at this early age. Forty-five percent of these children expressed a desire to lose weight, and 37 percent actually had tried!

Opting To Be a Thin Corpse

Consider another personal observation. The culinary bypass ("feeding without food") with nothingburgers and diet drinks can end up as a virtual addiction to aspartame products (see 10). These unfortunate women and men may then suffer severe withdrawal symptoms on attempted cessation. They include extreme tension, irritability, sweating, and a host of other features similar to those encountered when alcoholics go "cold turkey."

Given the current extent of bulimia and anorexia nervosa, this topic is no minor matter. In effect, it provides one answer to the question by Shelley Winters, "Where do you go to get anorexia?"

- A 29-year-old woman lost 80 pounds (!) over six months while on such a "diet."
- I saw a 32-year-old gaunt businesswoman in consultation for severe reactive hypoglycemia and depression. She was given advice about an appropriate diet and the avoidance of aspartame products. When this patient came three months later, she had lost an additional 17 pounds. My nurses were stunned by her cachectic appearance. When asked about the continued use of aspartame, she replied, "I'm still too fat. I'd prefer to be a thin corpse!"

> The patient returned two years later. She had suf-
> fered three witnessed epileptic convulsions. "Are
> you still drinking aspartame?" "Of course!"

On the subject of thin corpses, a recent film-farce bore the title The Pope Must Die. It was conceived in the long English tradition of cleric-bashing. The plot began with a mistake by a Vatican clerk that elevated a bumbling country priest to the papacy. In an attempt to defuse controversy, the film was promoted as The Pope Must Diet.

Sexual Innuendos

The TTO inevitably raises psychologic inferences about sexuality. A feature in The Palm Beach Post (June 14, 1991, p. D-1) titled, "Food Is To Women What Sex Is To Men,"illustrates the point.

Such reference to sexuality was reinforced in Eating by an actress seeking "a man who satisfies me as much as a baked potato." This bolt of revelation triggered prolonged hysterical laughter among several women in the theatre. In fact, I became uneasy about being summoned on an emergency "house call" for a stroke so precipitated.

Dieter's Boomerang

Women who indulge in aspartame products to lose or "control" their weight have paid a devilish price for what I term paradoxic weight gain. Indeed, some have gained 50 or more pounds! A 45-year-old woman described in Chapter 45 gained 75 (!) pounds. Here's another case history.

> A successful businesswoman experienced headaches,
> fluid retention, frequency and discomfort of urina-
> tion (day and night), severe lethargy, and depression
> while drinking aspartame beverages. But she was
> troubled even more by an unexpected gain of 50
> pounds. This is her own analysis:
>
> "I drank diet soda for the obvious reason — to avoid
> sugar and to avoid weight gain. The interesting thing
> is that I remember often thinking, 'My body knows
> that it is not sugar ...it's not fooled,' and I would go

looking for the real thing. In other words, it did not accomplish my goal in finding a sugar substitute. I cannot be sure it increased my craving for sugar...No matter how much I exercised, eating light, etc., I could not lose even a pound. All of the symptoms I mentioned above disappeared after discontinuing aspartame."

Others have alluded to this phenomenon, which could be likened to "friendly fire" military casualties.

- The September/October 1989 issue of Hippocrates carried a feature, "Why Diets Make You Fat."
- Noting the fact that the per capita consumption of artificial sweeteners rose from eight pounds in 1980 to about 20 pounds in 1989 (see 20), Skippy Harwood, food editor of The Palm Beach Daily News, despaired that no one had yet discovered a corresponding 20-pound per capita weight loss (October 18, 1990, p. 10).

I detailed the probable reasons for weight gain associated with the use of aspartame products in previous publications (see listing). They include (1) duping the brain's satiety center, (2) intense hunger of aggravated hypoglycemia, and (3) the deposition of fat following stimulation of insulin release by aspartame's amino acids, in tandem with the reflexive secretion of insulin after ingesting something sweet (see 16). (Insulin plays an essential role in the accumulation of fat.)

The word "repast" seems to have taken on new meaning with the advent of "lite" products. Its original connotation of a snack between meals (from the Latin "to feed again") is often transformed into "grand meal." This is consistent with the attitude of many dieters, "I eat because it puts food out of my mind."

Contributing Factors

One doesn't have to look long or far before uncovering the reasons for our society's dual affliction with fear of fat and the TTO.

A. Foremost are sophisticated <u>TV commercials and magazine ads</u> targeted at teenagers and young women. Even a cursory analysis of the front covers of their popular magazines reveals headlines preoccupied with fat, and the major focus upon growing smaller rather than wiser. Such emphasis culminates in the ill-advised, futile and possibly dangerous pursuit of an unreasonably low weight. Of course, products promoted as "sugar without the lumps" are understandably used.

B. <u>Clothes manufacturers</u> score really big in this department because of "thunderous thighs," especially with the female bathing suit. Even wise and confident Erma Bombeck felt like "a prisoner of my own thighs" while gazing at herself in a fitting room during such a shopping enterprise (<u>The Palm Beach Post</u> July 18, 1991, p. D-3). Acknowledging the salesperson's remark about the lack of perfect bodies, Erma confessed that women don't actually believe it.

C. The <u>emphasis upon "angularity" by many high priests of fashion</u>, rather than natural curvaceousness, merits comment. Coco Chanel, the famous designer, astutely observed that fashion is generated by men who basically hate women. Their motivation almost always involves the almighty buck. Translation: women who believe they are thin tend to buy more new clothes.

Ogden Nash quipped about yet another risk for men, above and beyond paying the bill, in <u>Curl Up and Diet</u>.

> "So I think it is very nice for ladies to be lithe and
> lissome,
> But not so much so that you cut yourself if you
> happen to embrace or kissome."*

D. Contemporaneous <u>models</u> who epitomize "thin and perfect" bodies are featured in ads promoting cigarettes and diet foods or beverages. This represents a recent phenomenon since food became abundant.

Erma Bombeck lashed out at these bored "cadavers"

---

* From <u>Verses From 1929 On.</u> ©1932, 1935 Ogden Nash. First appeared in <u>The New Yorker.</u> Reproduced with permission of Little, Brown and Company.

with vacant stares, tortured looks on their hollow faces, and exposed flesh-covered bones (The Palm Beach Post June 3, 1991, p. D-3). For good measure, she noted the high prices of some skirts that cover less space than a table napkin.

It has been estimated that only five percent of American women are capable of achieving the glorified model-thin look (The Miami Herald December 15, 1991, p.J-4).

There's the rub. This idealized figure largely reflects the unique genetic endowment for an abnormal body type rather than dieting, an eating disorder, or sleeping on a stretching machine at night.

To place the matter in clearer focus, models average 5-foot-10 and weigh 110 pounds. These figures contrast with the 5-foot-3 and 143 pounds of the "average" American female. The hurdles — and dangers — for her are magnified by the progressive reduction of body fat sought in contemporary models. Whereas the 1965 prototype was eight percent below the average body fat of most females (ranging from 22-26 percent), it declined to 23 percent by 1990.

E. Then there is the appeal of slim and highly photogenic female celebrities. They include movie stars (e.g., Katharine Hepburn), politicians (e.g., Representative Pat Schroeder), "first ladies" (e.g., Nancy Reagan), and even members of royal families (e.g., Diana, the Princess of Wales.) But there's often a joker in the former group. Some popular movie stars with relatively acceptable proportions have a "body double" who plays lovemaking and other nude scenes.

- A story about Carol Lombard intrigued me because my wife was named after this actress. It seems a studio boss regarded her as obese at 121 pounds, and lowered the boom. She proceeded to lose 10 pounds.
- The recent change in attitude toward the shapely endowment of women in noteworthy. For reference, artists in previous eras -- such as Rubens, Titian, and even during the Paleolithic age (e.g., the Venus of Willendorf) -- idealized the zaftig (plump, buxom) female figure.
- Prior to becoming a great actress, Sarah Bernhardt had been expected to be a "kept" women. This idea

fizzled, however, because she was too thin.
- The Sunday Times of London contrasted the differing views of Britains and Africans toward the svelte Princess Diana and the Duchess of York, her sister-in-law. It described the latter as "built more like an African woman with generous hips (and) broad backside" (The Miami Herald December 3, 1991, p. A-9). Whereas Diana was regarded as the epitome of beauty in the Western World, Africans regarded Sarah as the more beautiful. Diana was simply "too thin."

F. The fanaticism of peers and pressures exerted by mothers cannot be overestimated. Their two-fronted emphasis is well known in my area where many of "the rich and the famous" reside, especially during the winter. One hears ad nauseam the adage purportedly uttered in Palm Beach either by Barbara Hutton or the Duchess of Windsor, "You never can be too rich or too thin."

Also pity young women who display gusty appetites in public. According to Lafayette College researchers they are considered "unfeminine" by their peers, regardless of weight.

G. Even nursery rhymes can imprint this aversion early in life.
- "Jack Sprat could eat no fat, his wife could eat no lean."
- "What are little girls made of? Sugar and spice, and everything nice."

H. Then there are the weight-loss entrepreneurs. They include (1) manufacturers of "low calorie" products, (2) distributors of "revolutionary" dieting pills that can be taken without the inconvenience of reducing calories or increasing exercise, and (3) clinics and organizations offering "doctor-approved" or "doctor-supervised" weight loss programs.

I cannot leave this topic withour mentioning the plight of clients of such organizations and the frustration of many concerned owners of regional franchises. The latter recognize the problems of paradoxic weight gain and other reactions associated with the use of products containing aspartame by virtue of daily contact with individuals so afflicted. Their anger stems from two sources. First, the parent company incorporated aspartame (and often much salt) in its franchised products, notwithstanding protests by these dedicated business persons. Second, franchisees were forced to sign contracts stipulating that they not publicly criticize such products...or else!

I. Let us not overlook the authors and publishers of best-selling "diet books." The Wall Street Journal (June 3, 1991, p. A-1) correctly noted the tragedies stemming from this fusion of two great societal motivators: "love of money and fear of fat."

The list of diet fads and popular books promoting them is impressive. They include the Atkins diet, the Cambridge diet, the Stillman diet, and nutritional powders (with or without aspartame) for replacing meals. There also are tongue-in-cheek (or less-food-in-mouth) plans, such as (1) the I Love to Fly and It Shows diet (a person stays airborne for three months and consumes only airline food), (2) the Seafood Diet ("I see food and I eat it"), and (3) Eat-As-Much-As-You-Like Diet (but don't swallow it). All share these hitches: (1) a 95 percent failure rate over the long term in most studies, (2) the likelihood of binge eating, and (3) their self-defeating nature. The latter refers to the body's extraordinary efficiency in conserving every calorie as a result of repeated ("yo-yo") dieting.

Cartoonists have indulged in the manna provided by diet books and their shortcomings.

> Blondie presented Dagwood with a "fantastic" book
> on dieting (The Miami Herald November 10, 1991).
> She rejoiced over his enthusiasm...until finding him
> gorging on a classic Dagwood sandwich. When
> asked for an explanation, he pointed to a "loop hole"
> in the volume.

J. The medical and dietetic professions have climbed onto this bandwagon as alleged "fitness experts"...for hefty fees. Unfortunately, many pursued rigid attitudes of potential harm to patients.

The U.S. Department of Agriculture (USDA) issued a revision of suggested weights for adults in its 1990 Dietary Guidelines for Americans. Some physicians and nutritionists had a screaming fit, especially relative to the slightly increased weights suggested for persons over 35.

Why? They perceived bias "toward higher optimal ranges of weight" (<u>American Journal of Clinical Nutrition</u> 1991; 53:1102-1103).

> Surgeons and gynecologists continue to be admonished for using or recommending certain procedures in "the fertile field of fads and fashions." This phrase appeared in an editorial titled, "Victims of Fashion?", in the June 27, <u>1891</u> edition of <u>The Lancet</u>.

K. No discussion of this subject would be complete without brief mention of <u>two words</u> that can evoke panic greater than the mention of "Dracula." They are "foundations" and "cellulite."

- Erma Bombeck was impressed by the ability of <u>foundations</u>, also known as girdles, to withstand "a force greater than the Red Sea" (<u>The Palm Beach Post</u> May 23, 1991, p. D-3). Likening their effects to compressing garbage in trash compactors, or mulching dead foliage into compost, probably caused a run on diet sodas by Baby Boomers.

- Having done considerable research on obesity, I bashfully pleaded ignorance when I first heard the term <u>cellulite</u>. My embarrassment intensified with the perception that most women seemed to know all about this dreaded saboteur of thigh perfection. But there was a joker. They didn't.
  I subsequently learned that this spongy, cottage-cheese-like appearance isn't stubborn fat at all. Rather, it is a feature of feminine tissue conducive to dimpling at certain fatty areas. Furthermore, this benign frustration can't be "cured" by dieting, exercising, skin creams, spa therapies, or liposuction. In my opinion, the best treatment remains studious neglect.

Rays of Hope

I am glad to report a few positive developments that challenge the foregoing attitudes.

- A feature in the February 1990 issue of <u>Mademoiselle</u> Magazine was titled, "Big Girls Don't Cry: The Best Bodies Have 5 Lbs. Extra."

- The physician of a young woman with severe visual and neurologic symptoms attributed to aspartame products offered this startling encapsulated perspective: "I'd rather be fat than blind."

A Déjà Vu

Reading or watching many of the commercials for aspartame products aimed at weight-conscious women triggered an unpleasant memory. I kept recalling certain cigarette commercials during bygone decades that carried the stamp of approval of physicians, dietitians, the FDA, and even the American Medical Association.

A classic instance was the promo for Lucky Strikes® in which beautiful women allegedly kept their "youthful slenderness" by smoking this brand rather than eating sweets in between meals. It noted that 20,679 doctors regarded Luckies as less irritating to the throat than other brands.

With the benefit of enlightened hindsight, the question now should be asked relative to the risk of memory loss (see 19) and cancer (see 34): "Will history repeat itself if 'the brain' is substituted for 'the throat'?"

"The Wisdom of the Body"

Most physicians and scientists share my respect for "the wisdom of the body." It is a term coined many years ago by the famous physiologist Walter Cannon. In the present context, those much-loathed "five extra pounds on my thighs" (see above) represent important energy reserves. Indeed, these fat deposits probably constitute a major endowment by Nature for survival during periods of starvation. (This defense mechanism entails a "thrifty gene" involving insulin metabolism.)

The following letter from the Health Committee Chairman of the National Association To Aid Fat Americans illustrates the point. (I have no affiliation with this organization.)

"In checking through my files, I found that I'd run across your work before in the form of a 1970 article on 'Overlooked Dangers of Weight Reduction.' Any doctor who challenges the orthodox assumption that weight loss is the panacea for all the ills of the obese is even more remarkable...

"In the course of my researches, I have become convinced that weight reduction may carry many more dangers that those outlined in your article. In fact, there is hardly a single disease attributed to obesity which cannot be explained equally well as a consequence of chronic, repeated, or injudicious dieting. The greater the number of dieters in a given population, the more hazardous obesity appears to be. Thus the obese members of a popular diet club show a higher rate of illness than do the obese members of a more generalized population like Framingham...

"On the basis of these and other facts, it seems logical to propose that dieting may be a major cause —if not the major cause — of morbidity and mortality in the obese."

Some Medical Warnings
You'll just have to forgive me, Dear Reader, if this chapter reads like a continuation of my "sermon" aimed at concerned grandmothers (see 7).

I shall not detail all the gory details of tragedies resulting from reckless caloric restriction by young women seen in consultation. They are described in my medical publications — especially the precipitation of severe headache, heart disorders, thyroid gland problems, emotional disturbances, gallstones, and even multiple sclerosis.

Many reactors to "diet" products containing aspartame rue the day they embarked on this approach for weight loss. Some remember the exact date.

A 33-year-old analytical technician took four cans of an aspartame soda and 48 teaspoons of a "fasting supplement" daily to reduce. Describing the most significant reactions to these products that had affected her life, she wrote: "Unable to concentrate at work; missed a lot of work; became offensive to boss and co-workers, and also to family and friends...joints began aching, right ankle at first, and then both knees up to hips, shoulders, elbows and spine...could hardly walk at times." Even after eight months of abstinence, her "depression, dizziness, ringing in ears, nausea, memory loss and blurred vision" persisted.

Once again, Ogden Nash hit the nail on the head when he described the potential psychiatric devastation in <u>Curl Up and Diet</u>.

"Once upon a time there was a girl more beautiful and
    witty and charming than tongue can tell,
And she is now a dangerous raving maniac in a pad-
    ded cell,
And the first indication her friends and relatives had
    that she was mentally overwrought
Was one day when she said, I weigh a hundred and
    twenty-seven, which is exactly what I ought."*

The dangers are compounded by "diet pills" that can raise blood pressure and cause other side effects — including degenerative changes in certain areas of the brain involved with memory. But false or misleading advertising by this big business (e.g., "Eat all you want and still lose weight") continues to lure and ensnare the overweight...whether such excess poundage is real or illusory.

Morbid Details

Extreme voluntary weight loss can prove fatal for a variety of reasons. The most notable are alterations in heart function, and imbalances involving sodium, potassium and magnesium.

The same applies to <u>yo-yo dieting</u> with the use of aspartame products. This phenomenon refers to repeated cycles of weight loss and weight gain in millions of Americans, many of whom were not grossly

---

* From <u>Versus From 1929 On.</u> ©1932, 1935 Ogden Nash. First appeared in <u>The New Yorker</u>. Reproduced with permission by Little, Brown and Company.

overweight in the first place. Total mortality and mortality from coronary heart disease increased by about 50 percent among women with such fluctuating weights according to analysis of the extensive Framingham Heart Study statistics (New England Journal of Medicine, Volume 324, June 27, 1991, pp. 1839-1844). They nearly doubled in men.

Non-Morbid Details

The effect on mortality from a reduction of dietary fat poses another irony, especially for the legions of young adults now obsessed with PR-induced cholesterolophobia. A detailed projection of the mortality rates following reduced saturated fat intake from 37 percent to 30 percent suggests a measly two percent benefit in the "best case" scenario...equivalent to an increased life expectancy of only 3-4 months (Journal of the American Medical Association 1991; 265:3285-3291)! Moreover, such benefit accrues chiefly to persons over the age of 65.

Then there's the problem of diet-soda-guzzling fitaholics or exercise addicts. Such persons will continue vigorous and prolonged physical activity in spite of damage to tendons and ligaments, stress fractures, altered or absent menses, anemia, fatigue, and neuromuscular collapse.

The subcategories keep adding up, such as "thigh-thinner's thecitis." This refers to severe inflammation of the thigh muscles and hip joint after exercising with a spring-like device between the thighs in pursuit of losing "those five extra pounds." This phenomenon — also called exercise addiction or dependency — often reflects underlying severe depression and anxiety. These conditions are presumably relieved by the release of a natural opiate (beta-endorphin) from running or other rhythmic aerobic activity which induces a "runner's high." Hundreds of thousands of American women now regard exercise bulimia as a socially acceptable way to expend those perceived "excess" calories.

The Price of "That Mean Fit Look"

Professor Jan Smith was a highly-regarded teacher of geography at the University of North Texas. She also taught aerobics classes professionally. Under peer pressure to get "that mean fit look," she abandoned several basic aspects of sound nutrition, and began indulging in diet drinks and aspartame-laced chewing gum, popsicles and quicksnack foods. This became a "slender trap." A dramatic deterioration of her physical and mental health ensued — including a 30-pound weight gain, and the development of hyperthyroidism (Graves disease) (see 6).

There was a happy ending to this "case." Realizing that the only significant preceding change had been the use of aspartame products, Jan abstained from them. Gratifying and progressive improvement ensued...including loss of her weight gain.

A New China Syndrome

I have a suggestion for China's health professionals: read my publications on the potential hazards of aspartame to anticipate and avoid future problems.

This probably will strike the reader as incredulous. You are probably thinking, "O.K. I can understand the situation in developed countries. But China, home of the thin and the fortune cookie?"

Yes, China. I'm sure that manufacturers of aspartame and other "low calorie" products, as well as aerobics video producers, are carefully eyeing that market as rotundity replaces the thinness of famine and rationing. Just consider the overindulged obese offspring of one-child families...the officially approved limit.

A striking manifestation of this phenomenon is the enormous demand for special weight-loss camps (The Miami Herald September 26, 1991, p.B-6). They cost $110, the equivalent of 2-3 months income for an average worker. Overwrought parents are not likely to worry about ensuing aspartame disease if their teenager is repeatedly called a "big fat Chinese pig."

A Caveat

I must issue this pertinent social warning to any of my readers who become true believers: "Temper your pleas to friends and relatives about avoiding both aspartame products and severe caloric restriction. However well-intentioned, you risk generating contempt."

T. S. Eliot must have been aware of this unique twist of the human mind when he wrote

"This is what we must learn in relation to helping others. Because we care, we would like to
spare them our misfortune by warning them to
beware. However, owing to human nature,
rarely is this possible. Usually our advice is
scorned even to the point of ruining our friend-
ship if we should persist."

Other Guides For Weight Loss

One can hardly leaf through popular magazines without encountering articles and ads describing "virtually foolproof" ways of "shedding
those unwanted pounds." Most focus on low-calorie meals or products,
exercise, vitamins, and the discipline of "behavior modification."

Someone sent me an intriguing "Guide To Burning Off Calories." It
had been cut out of a magazine or bulletin. Unfortunately, neither the
name of the periodical nor the author was supplied. Since the message
is both humorous and relevant, I am reproducing it...accompanied by a
plea for information about its source to give proper credit.

	Calories
Passing the buck	25
Swallowing your pride	50
Beating around the bush	75
Jumping to conclusions	100
Climbing the walls	150
Throwing your weight around (depending on your weight).	50-300
Dragging your heels	100
Pushing your luck	250
Making mountains out of molehills	500

Luscious Earlobes

Finally, Dear Reader, allow me to reward you with this a Trivial
Pursuit® question. "What is the one part of a female's anatomy where
fat is actually desirable?" The answer: "Her earlobes!"

Believe it or not, some women seek transplants of fat from other areas to achieve "plump" earlobes. Several explanations for this current vogue come to mind.

- The "floppy" earlobes of Senator Paul Simon received national attention during his 1988 bid for the Presidency.
- A more convincing insight is the inference of larger earrings for luscious earlobes...a la "Diamonds Are a Girl's Best Friend."

Another paradox involves an addition, rather than subtraction, elected by weight-conscious women. The obsession of many nonobese females with small breasts led them to have silicone gel (a close chemical relative to Silly Putty) implanted in their breasts for this quasi-deformity. Most regarded micromastia as a disease! In point of fact, about two million (!) "victims" sought this "cure" until the FDA finally came to grips with the serious medical complications of silicone implants in 1992.

These women then encountered another formidable take-it-off challenge. Many plastic surgeons either refused to remove the silicone implants, or insisted upon receiving their fee (generally $3,000-$4,000) in advance.

25

"ASPARTAME IS EVERYWHERE"

The aspartame-is-everywhere theme has been repeatedly raised in discussions as well as their correspondence with reactors to aspartame products. Many bemoan the need for having to extend their antennas as far as possible in markets and drugstores to avoid the thousands of items that contain this chemical. Paradoxically, they include an array of "health products" such as laxatives, cold remedies and pain relievers.

A reporter for Central Maine's <u>Morning Sentinel</u> pressed the issue about stores being inundated with aspartame products. The light went on when I joked during the interview, "If it's not on your right, it's on your left" (July 31, 1991, p.2). A colleague had effectively used this analogy in a lecture to underscore the frequency of hemorrhoids in the general population.

"Everywhere" admittedly is a broad term. It also is pervasive on TV commercials, magazine ads and billboard promotions. Regarding the latter, Ogden Nash penned these prophetic lines in "Song of the Open Road" from <u>Many Long Years Ago</u>.

> "I think that I shall never see
> A billboard lovely as a tree.
> Indeed, unless the billboards fall
> I'll never see a tree at all."

The message for aspartame reactors is unequivocal: exposure to aspartame, as well as charity, begins at home. The life-threatening experience of Larry Wilson after swallowing an inadequately-labeled cold remedy that contained this chemical is recounted in Chapter 36.

<u>Restaurant Syndromes: Another Round</u>
Chapter 33 deals with the subject of medical syndromes. The "Chi-

---

*   From <u>Verses From 1919 On.</u> © 1932, 1935 Ogden Nash. *Reproduced with permission of Little, Brown and Company.*

nese restaurant syndrome" attributed to ingesting monosodium glutamate (MSG) is the granddaddy of "restaurant syndromes."

> Dr. Rose London, a Miami Beach internist, became a first-hand authority on "the onion roll syndrome" in her neighborhood (Cortlandt Forum March 1992, p. 61). A patron at the adjacent Gray's Inn required emergency care almost once a week after becoming pale, vomiting, and then collapsing with low blood pressure. The culprit turned out to be the restaurant's famous onion roll topped with onions, salt and Accent®. Such attacks resulted from sudden volume expansion within the stomach after rapid intake of large amounts of fluid and MSG.

Sulfite sensitivity is another. New products containing aspartame now have vastly expanded this category.

Make no mistake about it: aspartame reactors, as well as MSG and sulfite victims, take a real chance when they elect to dine out! They must question reassurances by the management that this sweetener is not present in the food or beverages being ordered, regardless of what labels may state (see 26).

> • A 43-year-old woman had frequent and excessive menstrual periods whenever whe consumed aspartame soft drinks. She wrote, "I am constantly irritated by people not informing me when a product containing aspartame is served. This has happened time and again at church suppers and friends' homes. I do not want it foisted on me!"

> • A 57-year-old woman attributed her violent headaches, dizziness, mental confusion, severe nausea and abdominal pain to aspartame products. She suffered repeated attacks after unknowingly being served aspartame, both here

and in Europe. During one such encounter, the ingredients of an "artificial sweetener" were not listed. She was later informed that the product <u>did</u> contain aspartame. As a result, this person now carries plastic bags in her purse in the event of unpredicted vomiting.

"57" Plus

The number of items containing aspartame exceeded the varieties of Heinz products almost from the outset. For interested readers, especially <u>Trivial Pursuit</u>® fans, I offer the fruits of my research on its "57" varieties.

Henry J. Heinz saw an advertisement for 21 varieties of shoes as he rode on New York City's elevated railway during 1896. This provided the stimulus to advertise his own products with a number...actually <u>any</u> number that could catch the public's eye. Heinz decided upon "57" — even though he already had produced more than 57 products. (Number One was his classic Baked Beans with Pork and Tomato Sauce.)

A Family Argument

Here's another vignette illustrating label confusion. It involved Betty, my fabulous mother-in-law. Being a diabetic, she was advised to avoid aspartame (as well as sugar) for reasons elaborated in my publications. Although this frustrated her sweet tooth, she faithfully adhered to my suggestion.

Carol accompanied her mother to a market during July 1990. To her astonishment, Betty placed several cans of a diet soda in her cart.

<u>Carol</u>: "Mother, that soda probably contains aspartame, which you shouldn't be using!"

<u>Betty</u>: "But it doesn't contain aspartame."

<u>Carol</u>: "How do you know?"

<u>Betty</u>: "The assistant manager read the label to me."

<u>Carol</u>: "And?"

<u>Betty</u>: "He told me it that it <u>specifically</u> stated there's no NutraSweet®."

<u>Carol</u> (reading the label): "This says it contains a 'non-nutritive sweetener.' The fellow probably thought 'non-nutritive' mean't 'no NutraSweet.'"

<u>Betty</u>: "Really?"

<u>Carol</u>: "And if you read the fine print on the label, it does say 'aspartame.'"

<u>Betty</u>: "What a shame! It was the only soda without aspartame I previously could find."

<u>Carol</u>: "I know, Mother. But that was before the company decided to shift from saccharin to aspartame ...like most of the others have done."

<u>Betty</u> (grinning): "I suppose this is the sweet revenge my doctor-son-in-law jokes about to me."

26

THE CHEMOPHOBIC DOCTOR

A best-seller book by Reverend Robert Fulghum avers that most of the important lessons in life should have been learned in kindergarten (see 27). To his list, I would add the saying, "Sticks and stones may break my bones, but names can never harm me."

I clearly recall the first occasion on which I found myself reciting this wise ditty relative to aspartame disease. It took place during October 1986 after my address to a national press conference in Washington, D.C. (see 23). I had done so at the urging of a consumer group, Aspartame Victims and Their Friends. Representatives of the manufacturer then asserted that I was afflicted with "chemophobia." The sticks-and-stones adage magically surfaced to conscious level at that point, and restrained my response.

I naturally knew the meanings of "claustrophobia," "agoraphobia," and other phobias. But this one required some real head-scratching! I finally concluded it meant "an aversion to all chemicals."

A concomitant thought consoled me — namely, I was in good company. I reflected on several esteemed physicians and scientists who had this epithet hurled at them for daring to challenge the safety of food additives, pesticides and other environmental contaminants...with anti-science and anti-intellectual Luddite implications. This attitude was later reinforced by the title of a symposium given at the 1990 national meeting of the American Chemical Society: "Chemophobia: Realities and Prescriptions."

Aspartame: A Synthetic Chemical

By way of brief review, <u>aspartame is a synthetic (or synthesized) compound</u>. It is NOT a natural or "organic" substance, as might be implied from certain ads (see below).

Aspartame contains three ingredients. Two are the amino acids phenylalanine and aspartic acid. For my scientifically-inclined readers, here's the formula of this chemical nectar.

ASPARTAME

ASPARTIC ACID   PHENYLALANINE   METHYL ESTER
(METHANOL)

Hearing or reading inferences that aspartame was "organic" or "natural" reflexively caused me to wince. Some cartoonists apparently had a similar reaction to this play of words. One depicted a corporate oil executive suggesting that sales of his gasoline could be boosted by advertising it as a "natural" cholesterol-free product that had been "organically produced."

"Building Blocks of Protein": Some Hitches

Several events that occur after ingesting the "building blocks of protein" in aspartame hardly can be regarded as normal physiology. One is the potentially large intake of its two amino acids. The other relates to consuming significant amounts of their uncommon dextro (D) forms or mirror-image stereoisomers. (The common levo or L form is physiologic; the D form is not.)

By way of brief explanation, meat and most protein-containing foods contain 4-5 percent phenylalanine in its L form. This contrasts with aspartame, which contains about 50 percent phenylalanine and 40 percent aspartic acid. Several biological problems (detailed in my scientific reports) ensue.

> • These amino acids are <u>rapidly</u> absorbed from the stomach and small intestine into the blood stream in amounts far greater than after the relatively slow digestion of conventional proteins.

- The brain is then flooded with phenylalanine. A problem arises because this occurs without the presence of other so-called neutral amino acids. As a consequence, the function may be altered of <u>very</u> important amino-acid-derived neurotransmitters involved in brain metabolism.

- The body may be unable to handle large amounts of the uncommon (D) stereoisomers of phenylalanine and aspartic acid — especially within the brain, nerves and eyes.

<u>A Case For Chemophobia</u>

I agree. The producers of aspartame products <u>should</u> be concerned when a "chemophobic" doctor with proper credentials tells consumers that there's considerable difference between ingesting amino acids in regular food or in this low-calorie synthetic sweetener. (Industry prefers the term "synthetic" rather than "artificial" or "fake.")

A distinction also must be made between chemophobia and "disinformation." Stated differently, "the facts" may not coincide with "the truth." This matter extends to "truth in endorsing" (see 41). In my opinion, ads that state the body doesn't know the difference between aspartame and beans, peas, grapes, milk or chicken may be misleading. The <u>Newsweek</u> June 30, 1986 assertion that the amino acids of aspartame are handled no differently than if they came from peaches, string beans or milk (in an ad) invites similar criticism.

<u>A Poison Is a Poison Is a Poison</u>

Now, be certain your seating is secure, Dear Reader.

Aspartame contains a <u>third</u> ingredient. Producers and their representatives have become masters at dodging the very mention of its name, especially on talk shows. They prefer to call this chemical "aspartic acid and phenylalanine as the methyl ester." In the body, however, the "methyl ester" translates into METHYL ALCOHOL...also known as METHANOL and WOOD ALCOHOL!!! The public is more familiar with methanol as an ingredient of fuels (such as Sterno®) and antifreeze. (The methyl ester is required for aspartame's taste since aspartyl-phenylalanine is tasteless.)

Just in case you didn't know, <u>methanol is a severe metabolic poison</u>. The dictionary defines "poison" as a substance having an inherent tendency to impair health or destroy life.

Significant amounts of methyl alcohol in its <u>free</u> form are <u>rarely</u> found in nature. The ingestion of relatively small amounts can result in blindness, other serious illnesses and even death.

> Methanol poisoning made headlines a few years ago when 25 persons died in Italy after they drank a table wine tainted with "only" 5.7 percent methyl alcohol (<u>The Philadelphia Inquirer</u> March 25, 1986, p. C-26).

For readers with a scientific bent, here are some facts about the free methanol released from aspartame.

- The digestion of aspartame yields at least <u>ten percent</u> methanol by weight! Specifically, the FDA calculated its presence in the aspartame molecule as 10.9 percent.

- One liter of an average aspartame-sweetened soft drink contains about 55 mg methanol. There is more, however, in certain beverages (such as orange sodas) to which greater amounts of aspartame are added in order to preserve the sweet taste.

Alice Recalled

Let's return to childhood, albeit not necessarily as far back as kindergarten.

When corporate representatives (see 17) defend the safety of aspartame in absolute terms, they tend to pass over the words "methyl alcohol" in a cavalier fashion. Or they may glibly state, "Why, there's more methyl alcohol in citrus juices than in aspartame!" (The essential facts were detailed in my previous writings.) Under these circumstances, I reflected on this "childhood" passage from <u>Alice in Wonderland</u>:

> "It is all very well to say 'Drink me,' but the wise little Alice was not going to do that in a hurry. 'No, I'll look first,' she said, 'and see whether it's marked <u>poison</u> or not'...She had never forgotten that if you drink very much from a bottle marked 'poison' it is almost certain to disagree with you, sooner or later."

An additional reference to Alice involves Lewis Carroll's other tale, <u>Through the Looking Glass</u>. I used this theme to provide perspective in medical matters. My article, "American Medicine Through Another Looking Glass" (<u>Journal of the Florida Medical Association</u> May, 1991, p. 276), focused on the importance of "perceptive and motivated doctors who are willing to view the numerous challenges confronting them through the looking glass of their own practice realities."

Alice's concern about "looking-glass milk" is highly appropriate to the subject at hand. Mention was made earlier of inherent biological problems associated with the D-form stereoisomers (or "wrong-handed" mirror images) of aspartame's two amino acids. As a result of her clear thinking, Alice has earned the Roberts' Gold Star for Literary Valor.

Identifying Aspartame Corpses

A remark by Mary Stoddard, President of the Aspartame Consumer Safety Network (see 36), intrigued me. In fact, my eyes nearly popped when she stated that a pathologist alleged he could tell if a decedent had consumed aspartame by measuring the body's formaldehyde concentration. (Methyl alcohol is indeed broken down to formaldehyde, and then to formic acid.) I could not refrain from asking, "Before or after the body was embalmed with formaldehyde?"

I also assigned Mary the task of hunting down any scientific paper on "sweet corpses" written by this or other pathologists.

Chemophobia and Labeling

The improper labeling of aspartame products enhanced my "chemophobic" attitude. This issue can provide critics of the FDA (see 42) with more ammunition.

The problem applies to other misbranded additives. For example, monosodium glutamate (MSG) (see 25) has been listed as "hydrolyzed vegetable protein" (HVP) — even when the label specifically states: "DOES NOT CONTAIN MSG" (The Wall Street Journal January 30, 1990, p. B-1). The consumer may not realize that MSG is present in HVP as a hidden ingredient.

A case in point is the near-fatal reaction suffered by Larry Taylor (see 36) after he swallowed a "cold plus" medication with a "lemony flavor." Failure to list aspartame as an ingredient in a clear manner nearly killed this known reactor.

New Lace: Let Them Eat Cake

Until recently, aspartame couldn't be used in cooking and baking. The NutraSweet Company can take credit for boldly tackling the problem of its deterioration on heating...to the delight of those minding the corporate cash register.

Innovative food technologists came up with a "heat protected" encapsulating process. A special polymer coating on each sweetener particle purportedly enables aspartame to survive higher temperatures and retain its sweetness. The production line and marketing eager-beavers then forged full-speed ahead. A health-watch column in The Palm Beach Post (March 8, 1990, p.D-13) carried this announcement:

> "Can you imagine a pound cake with almost a third fewer calories? Then visualize cakes laced with millions of gelatin micro-capsules that melt during baking and fill the dough with aspartame (NutraSweet), a non-carbohydrate sweetener. The cakes, puddings and other bake products should be in the supermarkets in three or four months."

The word "laced" caused this "chemophobic" to freeze as I

promptly recalled the title of a famous play, "Arsenic and Old Lace." Hopefully, psychologists will defer equating such an association with a personalized Rorschach test.

A Chemophobic Goes International

I received a letter in 1988 from Professor V. M. Stepanov of the Institute of Microbial Genetics in Moscow. Expressing considerable interest in my aspartame studies, he requested reprints. They were promptly sent. Over the next two years, however, I wondered about the reasons generating this unusual correspondence.

Several Soviet officials met Commissioner Carol Roberts during May 1990, and were duly impressed. For starters, she had salvaged the mission of Fazisi in the famous round-the-world Whitbread Race. This Russian yacht arrived in our area lacking both supplies and funds for the final leg of its journey because it was a private venture without state support. Ambassador Yuri Dubinin even made a special trip to meet Carol and to receive her proclamation of "Soviet-American Cooperation Day."

The Russian representatives recognized Carol's knowledge of both government and business, especially in joint-venture matters. In addition, one had learned about her husband's researches. As a result, they invited the Roberts to visit Russia, for which special visas were issued.

Since I would remain in Moscow for the several days Carol visited facilities in Georgia, I asked our hosts to arrange a conference with Professor Stepanov. He proved to be a charming person, as well as learned chemist. I presented him with a copy of my first book on aspartame which he had not been able to obtain. He expressed deep appreciation.

I naturally inquired about the Professor's interest in my work. His reply was unexpected. It seems that colleagues had developed a new method for the mass production of aspartame. When asked about its projected use, he stated, "They plan to add it to bread!" I thereupon urged him and any responsible officials to exercise extreme restraint before finalizing such a decision.

Two thoughts kept recurring during my return to the hotel. First, "I may have given the Russian people advice worth far more than most joint business ventures." Second, "I hope they don't spoil <u>maragena</u>" (the delicious Russian ice cream).

The Missing Research

I have expressed several major concerns about the safety of this chemical sweetener in my publications. They include (1) the absence of adequate long-term data on <u>humans</u> before aspartame was approved, and (2) the paucity of detailed studies on rodents and other species concerning tumors once the issue of brain cancer and other tumors (see 34) had been raised. To my knowledge, this information has not been published since the licensing of aspartame. Neither have the initial studies been repeated by corporate-neutral investigators, and formally presented.

Another reflection bears on the subject of chemophobia. Much of the effort and time I devoted to this subject might have been obviated if such studies had been done at some "Institute for Advanced Hindsight," as suggested by the cartoonist Sidney Harris.

Cartoon 26-2

The Missing Link?

Speaking of chemophobia, the brain has an apparent aversion to concoctions containing considerable phenylalanine and aspartic acid...as found in the chemical aspartame. This is relevant to the following items.

- The precipitation of confusion and memory loss in some persons after they ingest aspartame products (see 19).
- The large amount of aspartic acid, especially its D-form, in Alzheimer's disease plaques (see 28).
- The recent finding that a single amino acid mutation (involving substitution of phenylalanine for valine in amyloid precursor protein) appears to be the basis for at least one inherited type of Alzheimer's disease (Science 254:97-99, 1991).

These observations have potentially enormous significance in understanding the evolution, and possible prevention, of Alzheimer's disease. For instance, can the site of this "missense mutation" in the associated gene be protected?

Sidney Harris provided a remarkably perceptive cartoon in the September-October 1991 edition of American Scientist (p. 419). He depicted the drifting apart of "two amino acids" 3,562,398,027 years ago when life originated — and their "drifting together" 482,674,115 years later.

A Broad Psychological Brush

Considerable personal research on the toxicity of fluoride, as in fluoridated public water, led to another revelation. It was the remarkable overlap in attitudes by corporate advocates to critics of both fluoride and aspartame...namely, the inference that such vocal opponents ought to be considered as candidates for a straightjacket.

A case in point is the article by Dr. Michael W. Easley appearing in the summer 1985 edition of the Journal of Public Health Dentistry (pp. 133-141). It was titled, "The New Antifluoridationists: Who Are They and How Do They Operate?" Dr. Easley reviewed "the antifluoride personality" of devout intransigent leaders who purportedly regarded themselves as "saviors of their fellow men." He concurred with the assessments of others that they fell into two main groups of persons "capable of inciting intense fervor in others." One consisted of individuals "driven by factors of personal power, prestige or gain"; the other was "motivated by unconscious and powerful anxieties or hostilities." Furthermore, the personality profiles of antifluoridationists tended to focus on status deprivation, a lack of attachments, powerlessness, alienation, and individuals "marginal to the social, psychological, political and professional mainstream."

This tactical ploy invoked the subliminal 50-million-Frenchmen-can't-be-wrong argument. The ante, however, is more than double in both the aspartame and fluoride situations. Half the United States population currently consumes aspartame products. Similarly, 124 million persons in more than 7,800 communities now "receive the benefits of optimally adjusted fluoridated water." Question: could 120 million Americans be wrong if the warnings of credentialed "chemophobics" in each instance are valid? (I reviewed recent disturbing aspects of the toxicology of fluoride in the July 1992 edition of the Townsend Letter for Doctors [pp. 623-624].)

A Suit For Paranoia

I received an unusual call during February 1992 from a distressed Michigan physician. He phoned on my private line to indicate that his license was in jeopardy.

H.J.R.: "Why are you calling me about a licensure matter?"
Doctor: "It has to do with aspartame."
H.J.R.: "Really?"
Doctor: "Yes."
H.J.R.: "How did you get this number?"
Doctor: "Mary Stoddard of the Aspartame Consumer Safety Network gave it to me."
H.J.R.: "I see. Tell me the problem."
Doctor: "I may lose my license simply because I stated on television that aspartame products could cause serious reactions."
H.J.R.: "When was that?"
Doctor: "1984."

H.J.R.: "1984? Eight years ago?"
Doctor: "That's right."
H.J.R.: "Well, what's the basis of the state's case against you?"
Doctor: "In a nutshell, it claims any doctor who is paranoid enough to think that aspartame can cause medical problems shouldn't be practising medicine."
H.J.R.: "You must be kidding!"
Doctor: "Wish I was."

In effect, this physician was desperately seeking an "aspartamologist" with credentials who could provide expert testimony about the legitimacy of his observations.

A déjà vu surfaced. It involved several experienced doctors who had been summoned before state licensing boards for "administrative action" on charges of paranoia involving other "gray areas." The most notorious involved a general practitioner in upper New York State whose genuine interest in hypoglycemia had incurred the wrath of local endocrinologists. I reviewed the records of this dedicated physician. He lost his license, and was forced to sell vitamin products to survive!

Realizing the caller was in real trouble, I agreed to review the complaint and the facts before forming any opinion. Two considerations influenced this response. One was my curiosity about the matter. The other concerned a challenge to the First Amendment by prosecutors in cases involving the expression of honest observations and opinions, especially when it was suspected that an arbitrary legal suit had been provoked by self-serving corporate interests (see 11).

A Drug Is a Drug

Mention will be made of the conception of aspartame as a drug for possible use in treating peptic ulcer (p. 181). Such intent has been lost in the shuffle of "organic" and "natural" inferences.

According to written testimony dated January 11, 1974, Dr. J. Richard Crout, acting director of the FDA Bureau of Drugs, concurred with an FDA official that evaluation of aspartame's clinical safety as an artificial sweetener "be conducted under the investigational New *Drug* Regulations." (Italics supplied)

27

THE "MEDIA TERRORIST"

Another revelation occurred during my October 1986 press conference in Washington, D.C. In addition to being called a chemophobic (see 26), I was labeled "media terrorist." For months thereafter, my staff (especially my office nurse for nearly three decades) went into hysterics every time they heard this term.

For perspective, I reminded myself that if I was indeed a media terrorist, such activity had been done on my time and at my expense. By contrast, the braintrust that concocted this designation somehow failed to see the related irony of corporate expenditures. Some companies were willing to pay up to $700,000 for 30 seconds television time during Super Bowl football to advertise aspartame beverages whose safety had been challenged.

The Best Defense...

Speaking of football, Yours Truly got front-row and 50-yard-line exposure to the nitty-gritty of corporate self-defense. Part of it embodies the old saw, "Sometimes the best defense is a good offense."

One media theme — "Aspartame is the most thoroughly tested additive in history!" — kept reverberating over the airwaves, in medical publications, and even within the hallowed chambers of Congress. I even fantasized about a flag bearing this motto being placed alongside corporate officers and consultants during subsequent Senate hearings on aspartame, in full view of Old Glory.

A basic problem resides in the fact that offering dozens of "scientific" clinical studies doesn't prove safety when their corporate-approved protocols were flawed at the outset. I have commented at length on this "myth-stake" in previous publications. (Myths, of course, have played important roles in human behavior, such as the heroic tales spun by Homer and Rudyard Kipling.)

When I offered evidence in various forums about the shortcomings of such "thorough testing," corporate representatives predictably invoked certain diversionary tactics (see 17). A favorite focused upon the "anecdotal" nature of reported "idiosyncratic reactions." But there was a hitch. How could 6,000 (!) such "anecdotes" that had been <u>volunteered</u> by consumers to the FDA be ignored?

Nearly every fact I used to challenge this corporate-friendly assertion triggered the retort: "There you go again!" ...or something to that effect. As a result, I developed sympathy for those Republican presidential candidates whom Ronald Reagan had stung with this one-liner in the 1980 primary.

> Every action has a reaction, albeit long delayed. Whenever President Reagan exhibited confusion and memory loss during and after his two terms, especially concerning details of the Iran-Contra fiasco, I conjured up the likely thought of these former political opponents: "There he goes again!"

Sticks and Stones: Back to Kindergarten

I alluded to kindergarten as a societal and educational milestone in the preceding chapter, with particular reference to Robert Fulghum's delightful book, <u>All I Really Need To Know I Learned in Kindergarten: Uncommon Thoughts on Common Things</u> (Villard Books). After hearing the Seattle Unitarian minister deliver this simple message, Senator Dan Evans of Washington introduced it into the <u>Congressional Record</u>.

Two components of the Fulghum credo assert "Play fair" and "Say you're sorry when you hurt somebody." These themes even appeal to the medical profession.

> Dr. Joel Yager wrote an article in the Fall 1989 issue of <u>The Pharos</u> (publication of a national medical honor society) titled, "How Much Of What You Needed to Know in Medical School Should You Have Learned in Kindergarten?" For physicians, these tenets translate into integrity, honesty, compassion, fully informed consent, and altruistic concern for the public health.

Having been accused of being a "media terrorist" and a "chemophobic" after devoting four decades to primary-care medicine, I would offer these same "kindergarten lessons" in defense.

A Meeting of "Media Terrorists"

The University of North Texas sponsored a two-day symposium on the safety of aspartame products during November 1991. Several other "media terrorists" and I (Figure 27-1) were invited as speakers — at our own expense. We accepted this invitation notwithstanding the loss of several days from our practices or offices in such faraway places as Minnesota, Utah, Florida, Illinois and Washington, D.C. Barbara Mullarkey (see 23) referred to us as "Truth-Seekers Seven."

Photograph of the "Truth-Seekers Seven."
From left to right -- Professor Jan Smith, Barbara Mullarkey,
Dr. Paul Toft, Dr. H.J. Roberts, Gailon Totheroh,
James Turner, Esq., Mary Stoddard

Figure 27-1

This heterogeneous group assembled for a potluck dinner on November 7 before an evening open panel at the University. One of the participants soberly observed, "I suppose we <u>do</u> have to be careful. If someone <u>really</u> wanted to confront the leading anti-aspartame activists, it could be achieved at one fell swoop right there in this restaurant."

Another cloud enveloped this gathering. Professor Jan Smith, coordinator of the seminar, had just been fired for some dubious reason after having taught four years at the University. She was regarded as a great teacher, wonderful person and role model by most students who came in contact with her. Instead of backing off from the conviction that aspartame products had precipitated her affliction with Graves disease (see 24), Jan intensified her involvement as a "truth-seeker." I regarded her willingness to continue under these circumstances, in spite of considerable pressure, as a classic validation of the assertion, "One person with commitment can become a majority."

Brevity Favors Levity and Gravity

It <u>never</u> failed. A local television reporter or producer would request an urgent interview about aspartame for a "news feature." At the appointed time, he or she would boldly march into my office with an air usually only tolerated for visiting royalty. The interviewer then posed some provocative questions — for example, "Why does Dr. Roberts think aspartame causes (brain tumors) (Alzheimer's disease) (etc.)?" After a taping session lasting 10 to 30 minutes, this "media celebrity" and the accompanying cameraman hastily folded their tent and were off...at times with not so much as a "Thank you."

Maybe I'm an egomaniac. But having vested time and effort in providing such an interview, I wanted to see it. The final version was generally the same. My response rarely lasted more than 90 seconds. Moreover, it usually followed the <u>real</u> reason prompting the request: a recent statement allegedly confirming the "virtual" complete safety of aspartame products. The TV "bite," however, infrequently covered my reasons for having challenged such "scientific studies."

I have another bone of contention about these media personalities: their oft-unflattering introductions.

One "health editor" disrupted my schedule in an attempt to accommodate her. She started the 6 P.M. news interview by referring to me as "a local doctor who bills himself as an expert on the subject." This jab sorely tempted me to submit a fee for the interview.

Another aspect of such television editing is cause for greater concern. It reflects the anxiety of respected scholars and professional journalists who bemoan the severe shortcomings of commentaries by knowledgeable persons when condensed into a few minutes. I remain undecided as to whether this policy represents a time-is-money reflex or apprehension that the attention span of most viewers will be exceeded should the discussion last any longer. By denying the viewers "the rest of the story," producers shoot both the message and legitimate messengers.

28

SERENDIPITY AND THE EUREKA PHENOMENON

Many aspartame reactors experienced the Eureka ("I have it!") phenomenon. In a flash of perceptive insight, these individuals correctly linked their consumption of aspartame products to the cause or aggravation of unexplained complaints.

I was even more impressed when certain seasoned professionals related this hidden agenda of personal affliction because they shared the discipline of skeptical objectivity. These persons included health care workers, respected authors, and members of the media.

Serendipity

Dozens of aspartame reactors, or their relatives and friends, perceived such an association through serendipity. This term refers to "fortunate chance."

A recurrent scenario was the disappearance of symptoms while traveling in foreign countries where aspartame drinks were not available — only to suffer relapses after resuming them upon return. Such encounters even caused some who had recently visited so called "primitive" areas to reflect on which world was actually the more "civilized."

His and Hers

A young woman's fiance agonized with her during a year-long ordeal. She suffered severe headaches that had not been explained in spite of numerous consultations. Multiple therapies, including Percodan®, afforded no relief.

On one occasion, the couple happened to confuse their beverages. Her glass contained an aspartame-sweetened cola drink; his was a "regular" cola. It was the first time the fellow had tasted an aspartame beverage. After feeling "lousy," he urged her to avoid this product. His future bride's headaches diminished within a few days...and then disappeared. What a wonderful gift from a husband-to-be!

Correspondence From the Hypoglycemia Association, Inc.

Dorothy R. Schultz, devoted President of the Hypoglycemia Association, Inc., wrote me the following letter (reproduced with permission) on May 23, 1990. It contains another instance of serendipity.

> "At the same time your book arrived, I received a call from a man in Indiana who, at 39 years of age, was having a terrible struggle with hypoglycemia. He said his problems began when his doctor told him to give up sugar on the basis of a single sample of blood. He said it was a little high. It puzzled me that he would get so much worse after giving up sugar. It was much more than withdrawal symptoms. Further conversation revealed that he started using mints and chewing gum constantly, both containing aspartame. After reading your book, I called him back and told him what you said about the aspartame being absorbed very quickly through the mouth...The doctor gave him a 6-hour GTT (glucose tolerance test); he dropped to 40 in the 3rd hour. I guess the doctor didn't realize that he was going to go overboard on aspartame, and didn't warn him or didn't know enough about it to warn him."

I received this followup, dated August 16, 1991, from Dorothy.

> "I thought I had better check with the person to be sure that what I wrote was accurate, and also to see how he was doing since I last talked to him a year ago. So I called. He told me that he was working again (60 hours a week) and was about 90% improved ...He approved of what was written in my letter to you. He said it scares him when he thinks back about how sick he was."

"Networking By Accident"

Dorothy Schultz sent another letter one month later. It contained a title destined for inclusion in this chapter: "CHAINS OF EVENTS OR NETWORKING BY ACCIDENT."

She had read an article by a prominent dentist in a health-oriented magazine that focused on "what chewing gum manufacturers don't want you to know about what is in their products." The author made no mention of aspartame's detrimental effects. In response to Dorothy's letter on this subject, he stated that he was "glad to learn about Dr. H. J. Roberts' book on aspartame."

Sweet Discoveries

The stories of how each of the three leading sugar substitutes were accidentally found to taste sweet qualify as serendipity.

Aspartame. James M Schlatter, a scientist at the G. D. Searle & Company laboratory, was working on chemicals of possible value in the treatment of peptic ulcer. He heated aspartame in a flask of methanol during December 1965. Some of the mixture happened to land on the outside of the flask, and a bit got onto his fingers. He experienced "a very strong, sweet taste" when subsequently licking his finger to pick up a piece of paper. This was confirmed on retesting the original mixture.

Saccharin. Dr. Constantin Fahlberg synthesized o-benzosulfimide while working in a Johns Hopkins laboratory during 1879. Munching on bread that happened to be contaminated with this compound, he noted a striking sweet taste. It proved to be 300 or more times sweeter than table sugar (sucrose). He therefore named it saccharin.

Cyclamates. Dr. Michael Sveda, a young researcher at the University of Illinois, was studying various compounds that might reduce fever in 1937. He isolated the barium salt of N-cyclohexylsulfamic acid. Some tobacco shreds containing a bit of this chemical clung to his lips as he smoked, and induced an intensely sweet taste. The less toxic sodium salt, sodium cyclamate, was later developed and patented.

A Neurological Eureka for a Neurologist

I received the following self-explanatory letter from a female neurologist at the prestigious National Hospital for Nervous Diseases in London. She was researching new treatments for Parkinson's disease at the time.

Dear Dr. Roberts,

I was very interested to read your book about the possible dangers of the artificial sweetener, aspartame. I am writing to you now with yet another "anecdotal" report, concerning symptoms that I developed over the course of prolonged intake of aspartame-containing foods and drinks.

Aspartame was introduced into foods and drinks in the UK around 1983. Since then I have consumed large amounts of yogurts, soft cheeses, and drinks containing aspartame. On many days my entire intake consisted of aspartame-containing substances. I had a baby in April 1990, and 4 months later (around August 1990), I began to have problems using my hands and arms. I have always been physically strong, as I used to swim competitively, so I was surprised at the difficulties I was experiencing with simple things like picking up my son, undoing tight jars etc. Initially I put the problem down to over-use of my arms and didn't think much of it. However, I also noticed that my shoulder and forearm muscles had decreased in bulk.

Around Christmas 1990, I began to notice a cramp in my right leg whilst walking, especially in the cold. Then in January 1991 I noticed fasciculations [twitchings] in the small muscles of my hands, occurring many times a day. I began to notice them also in my shoulder muscles, forearms, calf muscles and thighs. At this stage, rightly or wrongly, I began to associate the problems I had using my arms (which had not improved) with the wasting in my shoulder and forearms, and finally with the fasciculations. I initially panicked thinking that I might have motor neurone disease. Of course, knowing how rare this was in someone of my age (I am 30), I began to think of alternative explanations.

I had not at the time come across your book, but as my work involves producing animal models of neurodegenerative diseases, I was aware of the toxicity of aspartate and glutamate to motor neurones. Early one morning, it occurred to me that my problem might be due to the huge amounts of aspartame (which I knew contained aspartate) that I'd been consuming over the past several years. It was then that I began to look through the literature for problems attributable to aspartame, and came across your book.

As you can imagine, training to be a neurologist and at the same time believing I was suffering from a neurological condition left me in a difficult position. I was worried that if I approached any of my neurological colleagues for an opinion, they would think I was mad!...Of my own accord, I immediately cut out all aspartate- and monosodium glutamate-containing substances from my diet. Within a month my symptoms and the frequency of my fasciculations had improved markedly.

I did, nevertheless, and with great embarrassment, approach a very sympathetic colleague (a Professor of Neurology at one of the London teaching hospitals) who examined me and could find no pathological signs. He also arranged for EMG and nerve conduction studies to be performed, but by the time the appointment for these came through my symptoms had almost completely resolved. (I had been off aspartame for nearly 3 months.) The studies were normal.

I adhere firmly to aspartame being responsible, as benign fasciculations would not explain either the difficulties I was having using my arms, or the disappearance of the fasciculations... Incidentally, I had also suffered increasingly from insomnia and migraine over the past few years, and these have almost disappeared since eliminating aspartame from my diet.

A Library Nugget

A young woman wrote me of her Eureka experience during a trip to the library.

Dear Dr. Roberts,

I am writing this letter to thank you for writing your book, <u>Aspartame (NutraSweet*):Is It Safe?</u> I happened upon it in the library. Because of the information contained in your book, I stopped drinking an aspartame cola and chewing gum a little over two weeks ago. Within the first 3 days, a rash I have had on my chest, back, thighs and face was nearly gone. It is now completely gone.

It is wonderful that you have the courage and perseverance to continue your research in the face of all the opposition from those with vested interests in products containing aspartame.

Another Revelation In the Library

A 49-year-old executive had been afflicted with epileptic attacks (both "temporal lobe" and grand mal seizures) for nine years. They persisted in spite of medical treatment. Other complaints included deterioration of vision, a constant ringing in the ears without demonstrable cause, and unexplained persistent itching.

I received a letter detailing these complaints in September 1991. It included this statement:

> "In early August, my wife discovered your book by happenstance at the library. Scanning its pages produced an 'AHA' experience: Aspartame <u>COULD</u> be the cause of my seizures because they <u>first appeared</u> when I started using aspartame. I immediately removed all aspartame from my diet...One clear thing DID happen DIRECTLY — the hand symptoms (itching) went away."

One In Two Million

Metropolitan Miami has two million inhabitants, give or take several thousand. This statistic provides background for a remarkable its-a-small-world anecdote.

My wife, a Palm Beach County Commissioner (see 6), served in various related capacities. One was as Chairperson of the Tri-County Commuter Rail Authority linking Palm Beach, Broward and Dade Counties. Under her leadership, "Tri-Rail" evidenced dramatic and successful growth.

The efforts of Priscilla Perry, the lobbyist for Tri-Rail, impressed Carol in its attempt to get a $45 million federal grant for further expansion and additional trains.

Carol made a suggestion during August 1991. "I think you'd enjoy coming for dinner with a charming person who is Tri-Rail's lobbyist. Her name is Priscilla Perry, and she lives in Miami. If it's OK with you, we'll meet Priscilla and her husband half way in Fort Lauderdale. Incidentally, I have absolutely no idea what he does."

We met at the Down Under Restaurant two weeks later. Carol introduced Priscilla's spouse as Eugene. I assumed that the last name of this distinguished-looking chap was Perry. Priscilla and Eugene proved to be brilliant persons and animated conversationalists.

The conversation centered initially about Priscilla's work and the fascinating story of her evolution as a lobbyist. The discussion then turned to other general topics. I finally asked Eugene about his activities.

Eugene: "I'm President of the Center for Health Technologies."

H.J.R.: "What is its major focus?"

Eugene: "Biomedical research."

H.J.R.: "And what's your role, Gene?"

Eugene: "I'm a Ph.D. chemist."

H.J.R.: "What field in particular?"

Eugene: "I'm involved with amino acids."

H.J.R.: "That's amazing."

Eugene: "What do you mean?"

H.J.R.: "Well, I'm sure this is going to sound strange, coming from a practising physician, but I happen to have <u>considerable</u> interest in several amino acids."

Eugene: "Which ones?"

H.J.R.(with overt reluctance): "I really shouldn't be bor-

ing you and Priscilla with such details at dinner."

Eugene: "It's perfectly alright! Tell me, which amino
acids?"

H.J.R.: "Phenylalanine and aspartic acid, including
their stereoisomers."

Eugene: "How come?"

H.J.R.: "Chiefly because of my involvement in problems with aspartame. You know, the artificial
sweetener? In fact, I've had several interesting
discussions with a peptide chemist in Southern
California named Jeffrey Bada."

Eugene: "Brace yourself. That's _my_ field and he's _my_
buddy!"

H.J.R.: "I don't understand."

Eugene: "I worked with Jeff. Here's my card." (It read
"Eugene H. Man, Ph.D.")

H.J.R.: "_You're_ the Man who coauthored the review on
D-amino acids?"

Eugene: "The same."

H.J.R.: "I can't believe it! In fact, I was planning to call
you after Jeff told me about your association
with the University of Miami."

Eugene: "Why?"

H.J.R.: "I cited several of your papers as references, and
thought you might have some recent informa-
tion about D-aspartic acid concentrations in
brain and nerve tissue."

The discussion got more technical — and exciting — over the next
half hour as we compared notes. It encompassed aspartic acid concentrations in tendons, Alzheimer plaques, and the brain of rats fed alumi-
num.

All the while, I expressed pangs of social conscience, and apologized
to Carol and Priscilla for having engaged in such "business" talk. Both
were delighted over this extraordinary coincidence because it gave
them the opportunity of engaging in their own specialized shop talk.

29

SOCIAL HISTORIES

I have documented the convincing severe reactions to aspartame products — including headaches, depression, visual problems, confusion, memory loss and menstrual disturbances — suffered by hundreds of business and professional persons, and their subsequent improvement merely by avoiding them.

The frequency of informal "curbstone consultations" about these matters increased dramatically whenever the print and electronics media featured my studies or an interview. In fact, hardly a week now passes without some acquaintance raising a related issue. Nurses and secretaries at my two hospitals who volunteer similar problems also fall in this category.

"Casual" Questions

Casual questions asked by friends represent yet another variation on this theme. The third degree wasn't required to infer that they, a family member, or some "significant other" person had probably experienced an aspartame-related reaction.

> The wife of a prominent elected official sat next to me at a dinner party. She asked, "Could aspartame cause palpitations?" I replied in the affirmative, and mentioned several recent instances. As we parted an hour later, she said, "Thank you so much for telling me about your studies on aspartame...especially the palpitations."

An "Old Buddy" Tale

Other persons went out of their way to express appreciation for the dramatic improvement of previously-undiagnosed disorders once they avoided aspartame products.

The wife of an executive approached me at the annual meeting of a major local organization. I had known this woman and her family for 30 years. She stated, "I'm most grateful for your television interview about aspartame reactions two weeks ago. I stopped drinking them, and all my symptoms vanished! But my husband is even more grateful. Let <u>him</u> tell you." This no-nonsense businessman then elaborated.

"I didn't know what was happening to me. Come the middle of the afternoon, I felt terribly sleepy and couldn't think straight. It was like my brain was being lifted. I began to worry about developing early senility. My wife then told me about her reactions to aspartame drinks. The sleepiness and confusion disappeared right after I stopped them, and haven't returned. Thanks, old buddy!"

A Chance Airport Meeting

A chance meeting on December 19, 1989 proved extraordinary. It relates to eye complications associated with the heavy consumption of aspartame-sweetened beverages.

I arrived at the Palm Beach International Airport coincidentally with two acquaintances. I was en route to Virginia Beach at the request of <u>The 700 Club</u> for a television interview on my soon-to-be-released book on aspartame reactions (see 14).

The wife, a distinguished Ph.D.-psychologist, wore a patch over her left eye. In response to my expressed concern, she stated that a hemorrhage had occurred for which no cause could be found. In fact, she had just consulted a nationally-known ophthalmologist who was equally perplexed.

I reflexively asked about the use of aspartame-containing products. Her husband blanched. "Why, she drinks at least 12 of those d - - - diet colas every day! I've been after her to stop because I think they could be harmful. But she simply won't listen!"

To remain silent risked a guilty conscience. I thereupon opened the advance copy of my book to a page describing a comparable case. The psychologist read it, hesitated, and replied, "OK, I'm convinced!"

Freud and Taste Buds

The wife of a prominent lawyer-acquaintance developed severe headaches, confusion and other symptoms after increasing her consumption of diet sodas. Although she suspected that they were the cause, a "possible psychological hangup" kept surfacing.

It seems that my friend's daughter once had a date with the son of an executive of G. D. Searle & Co. (the original United States aspartame manufacturer). The lad then took out another girl the same night. This infuriated the mother. So whenever "the aspartame connection" was raised, she asked herself, "Could this be a Freudian slip of my taste buds?"

A Friday-the-13th Encounter

I was returning from Boston on April 13, 1990...a Friday. The senior pathologist at one of my hospitals happened to be on the flight.

Pathologist: "By the way, I meant to tell you about my own problem with aspartame."

H.J.R.: "Oh?"

Pathologist: "Yes, it affected my vision."

H.J.R.: "In what way?"

Pathologist: "Things began to look very hazy...like it was cloudy outside when the sun was actually shining."

H.J.R.: "How did you relate it to aspartame?"

Pathologist: "Well, I did a lot of thinking when confronted with the possibility of cataract surgery. All I could recall doing any different was drinking more diet sodas. So I quit them."

H.J.R.: "And?"

Pathologist: "My vision cleared almost completely in a few days."

H.J.R.: "Any other symptoms?"

Pathologist: "Not really. But there's more proof."

H.J.R.: "Tell me."

Pathologist: "I naturally wondered if the matter might have been only a coincidence. So I resumed drinking diet sodas."

H.J.R.: "What happened?"

Pathologist: "My vision got <u>real</u> foggy in just a few hours. Believe me, I'm a true believer now!"

Judging a Judge

Being the husband of a county commissioner, I receive "spouse" invitations to affairs that few physicians attend... especially parties hosted by trial attorneys. One was a western-dress event at a local ranch sponsored by a prominent law firm. Several senior judges and their wives also attended. One approached me.

Judge: "Why, hello, Dr. Roberts! It's been a long time."

H.J.R.: "My pleasure to see you and your wife again, Judge."

Judge: "You know, I first learned about how hypoglycemia can affect behavior from you."

H.J.R.: "Really? I don't remember the circumstances?"

Judge: "The case involved a prominent merchant's wife who was apprehended for allegedly driving under the influence. You might not recall since it must have occurred at least 25 years ago."

H.J.R.: "Now that you mention it, I certainly do! In fact, that episode occurred in the middle of the afternoon."

Judge: "Right!"

H.J.R.: "And I presented my studies comparing the results of her morning and afternoon glucose tolerance tests."

Judge: "Exactly!"

H.J.R.: "It so happens that there's a figure comparing those tests in one of my books. It dramatically shows that her blood sugar level done three hours after the test began was 30 points lower in the afternoon than in the morning."

Judge: "I also was impressed by the effects of sugar on behavior and brain function in patients with hypoglycemia."

H.J.R.: "Believe it or not, Judge, the effects of the artificial sweetener aspartame can be just as bad."

Judge (surprised): "What effects?"

H.J.R.: "Oh — like headaches, confusion and memory loss."

Carol (my wife interjecting):   "You're a close friend of _____
(a top executive), aren't you?"

Judge: "Yes."

Carol: "Just ask <u>him</u> about his problems while he was drinking diet colas. Since he's not a patient of my husband, I can tell you that he actually thought he was developing Alzheimer's disease. Why, there were times he couldn't remember whom he had just dialed on the phone."

Judge: "Really?"

Carol: "You bet!   And as soon as I mentioned some of my husband's studies, he stopped those sodas, with prompt improvement of his memory."

Judge: "How could he be sure?"

Carol: "Easy. The same thing happened when he resumed the diet drink. But that's an experiment he tried only once."

Judge's Wife (to her husband): "Could that be why <u>you're</u> having those headaches and spells of confusion?

Judge (reflecting): "Maybe."

Wife: "Just how many cans of diet cola do you actually drink a day?"

Judge: "Between four and five."

H.J.R.: "Judge, it's obviously not proper to give you any medical advice. And I'm not here to sell my book. But here's the title anyway. You may find you have a lot in common with many of the patients reported... including your lawyer-friend!"

Lightening Strikes Twice

At least one colleague is grateful for my discussion about Aspartame Disease — Dr. Nicholas Petkas, the talented creator of many cartoons in this book.

Nick seconded the visual problems experienced by our pathologist-friend (p. 192). He volunteered this observation: "I rarely get ocular migraines any more since avoiding diet drinks. But before you clued me onto the matter, I was suffering from these attacks [often referred to as 'chained lightening' by patients] once or twice a week."

30

ALL IN THE FAMILY

An unexpected revelation surfaced during my researches. When persons with reactions to aspartame products were questioned <u>specifically</u> about the matter, I found that comparable reactions had occurred in at least <u>one-fifth</u> of their families! The current record is <u>seven</u> persons from the same family so afflicted.

There's an important "bottom line": persons who suffer reactions to aspartame products should relay this information to their relatives. In several instances, the first awareness of an "aspartame connection" developed during a family reunion. Ogden Nash appropriately observed

"Our daily diet grows odder and odder —
It's a wise child that knows its fodder."*

Instances of severe reactions to aspartame products among mothers and their children (usually daughters) will be described. The combination of case histories and the inference about other abnormal maternal influences, suggested by the title of this book, caused my jaw to drop on hearing two words uttered in <u>The Lion In Winter</u> by James Goldman: "Unnatural, Mommy?"

<u>A Family History</u>

A <u>homemaker</u> developed depression, recurrent pains in the head, palpitations, and visual difficulty after drinking aspartame-containing colas. She described her eyes as "getting weaker...as though they had a thumb on them, but when I rubbed them, it didn't help." All her symptoms vanished four days after discontinuing aspartame pop.

A <u>sister</u> had been a "very happy, outgoing person." She then suffered multiple reactions to aspartame-containing beverages — chiefly severe depression and unexplained pain in the arms. She reverted

---

* From <u>Face Is Familiar</u>. © 1933 Ogden Nash. Reproduced with permission of Little, Brown and Company.

"back to her happy normal self" within several days after stopping them, and had no more arm pain.

Another sister also experienced intense depression. "She didn't care if she got up in the morning, and had little interest in anything." On learning about the reactions of her two sisters to aspartame products, she avoided them...and felt fine within three days.

A non-smoking niece had unexplained chest pain that promptly subsided after abstaining from aspartame products.

Cast In the Same Mold

Another interesting phenomenon is pertinent. It concerns the similarity of reactions to aspartame products — that is, primarily affecting the same organ — in multiple family members.

- A woman and two close relatives developed severe diarrhea after drinking variable amounts of aspartame-containing cola.
- Each of two 40-year-old identical twin sisters experienced severe abdominal pain as the chief adverse reaction to aspartame products.
- The most prominent reaction to aspartame products of a 62-year-old female was "immediate difficulty in swallowing." She wrote, "I was eating a cereal and did not know it contained aspartame. My throat became paralyzed and I could not swallow. My daughter asked if I had checked for aspartame. When I did, that's when I realized that I was using it." The daughter had experienced "throat paralysis" as her predominant major reaction to aspartame products.
- Each of two women with aspartame-related seizures had several children who also were afflicted with convulsions. Both mothers consumed aspartame during pregnancy and while breastfeeding.
- The menstrual periods ceased in a 45-year-old interior designer two weeks after she began using a diet cola and "diet food meals." Concomitantly, her two

teenage daughters stopped menstruating. Three months later, aspartame products came under suspicion as the cause. Normal menstrual periods returned <u>in all three</u> within one month after avoiding aspartame!

<u>Double-Duty Questionnaires: A Study in Cost-Consciousness</u>
Two or three members of several families filled out the <u>same</u> questionnaire form, even though more would have been sent free if requested.

A 22-year-old Air Force officer and his 24-year-old sister used the same form to list their reactions to aspartame products. He used <u>blue</u> ink; she used <u>red</u> ink. Both experienced severe headaches and extreme irritability "immediately after usage."

<u>A Sweet Niece</u>
My niece Rebecca is a lovely young lady. Her charm, intellect and proficiency as a gymnast were a source of delight and pride for the entire family.

Rebecca's parents gave a party on the occasion of her confirmation. Among those asked to light the ceremonial candles were "My Aunt Carol, and my famous uncle who just wrote a book about the nasty things aspartame can do to you." Recoiling from momentary shock, I realized that "Becky" had scored an important point...and one for which some of her captive peers might thank her in years to come.

<u>The Physician Connection</u>
I'll be frank. Few of my medical colleagues <u>really</u> took my warnings about reactions to aspartame products seriously. There were two notable exceptions: physicians who had <u>personally</u> suffered a reaction (see 29), and those who witnessed it in a spouse or child.

A cartoon by William Boserman in <u>Medical Tribune</u> reminded me of these encounters.

"Dr. Turner has a patient with the same thing his wife had. What did you do for her?"

Cartoon 30-1

Marriage On the Sweetened Rocks

Some of the more frequent reactions to aspartame products (see Overview) are hardly conducive to marital harmony. They include marked irritability, severe anxiety, extreme fatigue, personality changes, phobias, depression, insomnia, and intense drowsiness.

Geraldo's Show of July 24, 1989 underscored the ability of both sugar-induced hypoglycemia and aspartame reactions to torpedo marriages.

> A chap related his wife's insomnia and dramatic personality deterioration six months after their marriage. He attributed them to radical change in her eating habits. A divorce ensued. When he read JoAnn Cutter Friedrich's book, "The Pre-Menstrual Syndrome: How to Tame the Shrew in You," he communicated with his "ex" about her diet. The gratifying response that ensued led to a resumption of their relationship.
>
> A psychologist who specialized in food addictive disorders told this audience that the last thing sought by most persons who become very tense on caffeine, sugar and aspartame is an intimate emotional relationship. JoAnn Cutter Friedrich responded, "If you don't have the right brain chemicals, then you're going to respond inappropriately." I hardly consider the ingredients of aspartame (Chapter 26) "the right brain chemicals."

Nurture...Not Nature

It is easy to ascribe the occurrence of a disorder among multiple family members to the combined influences of "nature" and "nurture."

On the other hand, I have encountered several dramatic instances in which an environmental factor — specifically, the intake of aspartame products — was clearly operative among legal relatives who were not genetically related. Two genetically unrelated step-sisters who developed hyperthyroidism (Graves disease) are described in Chapter 6. This raised a similar inference for President and Mrs. George Bush, each of whom developed this condition and had consumed aspartame (see 6).

Relatively Speaking

I was scheduled to address the Section on Neurology of the Southern Medical Association at its annual meeting in Atlanta on November 17, 1991. The subject: "Neurologic Reactions to Aspartame Products: A Contemporary Epidemic."

I attended a luncheon-seminar several hours before giving this lecture. A physician from Pine Bluff (Arkansas) and his wife sat next to me.

Wife (looking at my name tag): "Are you the famous Dr. Roberts from West Palm Beach?"

H.J.R.: "Well, I'm one of several Dr. Roberts in the area."

Wife: "Do you happen to know the Pallots in Miami?"

H.J.R.: "Of course! My wife's mother is a Pallot."

Wife: "Then you must know Norman Pallot?"

H.J.R.: "Certainly! Why do you ask?"

Wife: "I'm his wife's sister!"

H.J.R.: "Isn't it a small world?"

Wife: "Aren't you also the doctor who has written about medical problems caused by diet sodas?

H.J.R.: "Yes."

Wife: "I declare! You certainly made true believers of Norman and Ann. Why, they won't even allow the stuff in their home!"

H.J.R.: "It's reassuring to know that some of my relatives actually listen to me."

A Researcher and His Spouse

Dr. Richard Wurtman is a respected researcher at the Massachusetts Institute of Technology. He also was among the first physicians to correlate the occurrence of seizures (convulsions, epilepsy) with consumption of aspartame-sweetened products. His subsequent investigations in animals indicated that aspartame can lower the threshold for experimental-induced seizures.

On a personal level, Dr. Wurtman wrote the Wednesday Journal (June 21, 1988) that he enjoyed diet sodas late in the afternoon, particularly for the caffeine effect. Not so his wife, however, who "...tries to avoid aspartame because she believes it gives her headaches."

I hope that Richard will see fit to publish the results of double-blind studies on his spouse in view of the remarkable convenience for performing them at home. If no grant monies can be obtained, I will help to defray the costs of the aspartame-sweetened beverages and foods required.

31

NEWSPAPER CLIPPERS

Some of my patients are avid clippers, as evidenced by nicely-excised excerpts received from newspapers and magazines over several decades. This stream increased dramatically following my interest in reactions to aspartame products.

I <u>always</u> respond to such thoughtfulness with an expression of thanks, especially when not previously aware of some important point contained in these media nuggets.

Better Late Than Never

The occasional long delay in receiving such goodies from my newsclipping benefactors didn't bother me.

> A patient wrote me on February 16, 1990: "Thought you might be interested in this clip." An enclosed lengthy feature dealt with my initial press conference concerning the first 100 aspartame reactors in my series. It had appeared in <u>The Chronicle-Telegram</u> of Elyria (Ohio) as a feature titled, "Aspartame Not Such a Sweet Ride?" The date: August 3, 1986.

To Use or Not to Use?

I was tempted to incorporate a number of clipped gems in my manuscripts, but generally hesitated. A valuable therapeutic technique, tincture of time, usually resolved the dilemma. I would let some juicy literary morsel simmer on the back burner of my mind for weeks or months, and then ask two questions. First, "Is it appropriate?" Second, "Where should I put it?"

An offering by Jac Wilder Versteeg in the May 3, 1991 edition of <u>The Palm Beach Post</u> (p. A-10) illustrates this point...ultimately ending up here. The title read, "Playing Chicken: The Lite Thing to Do." It was the subtitle, however, that intrigued me: "What's a 'Better-For-You'

Product? Something That Just Kills You Slower?" My attention became riveted on Jac's discussion of sugar-free sodas.

> "Better for you than tooth-rotting sodas. On the other hand, sugar-filled sodas are better for you than chemical-laden sodas."*

The last paragraph of this clipped article hit home.

> "What's unsaid about these better-for-you products is that none is a good-for-you product. They're all just variations on another product — Reality Lite. Like all junk food, it's easy to fill up on it. And the manufacturer is betting you won't read the ingredients too closely. I try not to. Scared what I'll find. That's right. I'm chicken."*

The Doctor, The Clipper

There's a personal aspect to this clipping matter. Years ago, I developed the strange habit of reading newspapers and scientific journals with a razor in hand. My proficiency was evidenced by the ability to cut out an article without hardly indenting the underlying page. This "talent" repeatedly impressed my staff. Although familiar with the novel, The Razor's Edge, they obligingly avoided mentioning its title.

This habit proved costly, however. To keep peace in my family over the disturbing "windows" so created, I opted for double subscriptions to The Miami Herald and The Palm Beach Post. The brownie points received for contributing so generously to our paper-recycling effort reduced the anguish associated with such added outflow of bucks.

I'll confess that this pastime was not limited to medical features. My equal-opportunity blade also extracted some political items. As they mounted, many had to be "filed by pile."

There was a downside to this effort in the form of my wife's verbal jab, "What in the world are you ever going to do with all those pieces of

---

* © 1991 The Palm Beach Post. Reproduced with permission.

paper?" Shrugging my shoulders, I uttered, "They're for future reference, Dear."

Such criticism on the home front became dramatically curtailed during April 1990. Carol was running for re-election to the Palm Beach County Commission. She faced unexpected opposition in her primary race by the former mayor of a nearby city. My Favorite Politician thereupon mentioned the need for visiting "the morgue" at The Palm Beach Post.

> H.J.R.: "I really don't think that's necessary."
> Carol: "What do you mean?"
> H.J.R.: "Well, it so happens that I have a whole folder on your opponent."
> Carol (in disbelief): "You do?"
> H.J.R. (several minutes later): "Here's a batch of clippings about some of that ex-mayor's hanky-panky activities while in office."

P.S. Carol handily won this primary against two opponents without the need for further resort to my "future reference" files.

32

THE ULTIMATE QUESTION

I have cautioned patients and the public about the use of sugar and concentrated sweets for more than three decades. Such advice also appeared in dozens of publications dealing with my researches in hypoglycemia ("low blood sugar attacks") and diabetes mellitus.

This counsel did not represent a nefarious obsession aimed at making the lives of persons with a sweet tooth miserable. Rather, it seemed the logical conclusion of many studies suggesting such prudence. (Admittedly, some respected diabetologists recommend larger amounts of complex carbohydrate and a Mary Poppins' bit of sugar in the diet.)

My increasing concern about the adverse effects of aspartame products in patients with hypoglycemia and diabetes mellitus led to a shift in emphasis. Indeed, I found myself between the proverbial rock and hard place when asked, "But, Doctor, which is worse for me...sugar or aspartame?"

The proper answer required considerable soul-searching. When now cornered on the matter by a diabetic patient or talk show host, I generally reply, "If push comes to shove, a bit of sugar is probably preferable." In the case of patients subject to reactive hypoglycemia, I add, "...as long as you eat balanced meals, and take a small snack in the middle of the afternoon and at bedtime."

Cartoon 32-1

33

MEDICAL SYNDROMES OR "WHAT'S LEFT IS WHAT YOU'VE GOT"

I emphasized my attempt to minimize "medicalese" in the Preface. On the other hand, several terms such as "syndrome" cannot be dodged, especially in view of the belief that "aspartame disease" represents a <u>brand new</u> and important syndrome. Webster's <u>New Collegiate Dictionary</u> defines syndrome as "a group of signs and symptoms that occur together and characterize a disease."

Physicians understandably seek such clues and perspectives when they encounter strange disorders in practice. I can vouch for the credibility of this assertion as author of <u>Difficult Diagnosis: A Guide to the Interpretation of Obscure Illness</u> (W. B. Saunders Company, 1958), also published in Spanish and Italian editions.

Cartoonists like to capitalize on the diagnostic uncertainty that besets doctors and their consultants. This was nicely illustrated by William H. Boserman in the September 24, 1986 issue of <u>Medical Tribune</u>.

"*The laboratory tests were negative and her psychiatrist said it isn't hypochondria. Whatever is left is what she's got.*"

Cartoon 33-1

© 1986, <u>Medical Tribune</u>. *Reproduced with permission of The Medical Tribune Group and Mr. William H. Boserman.*

Sir William Osler was absolutely right! He urged young physicians always to listen to the patient for two perfectly good reasons: he or she often can reveal what's wrong, and why it happened.

The frequency with which patients ultimately made the correct diagnosis of aspartame disease has been remarkable. Here are a few instances.

- "I suffered for months with an invisible rash, scratching to the point that it made people around me uncomfortable...I discovered through elimination of food products that my problem was aspartame. When I went to the dermatologist, she told me it was impossible. I knew she was wrong because I drank a diet soda, and started to scratch within 15 minutes."
- "From my personal experience, I feel aspartame's effects almost immediately when I drink it. I could drink a soda with aspartame at 1 P.M., and have a headache by 1:30 P.M."
- "The diagnosis of this reaction (seizures) to aspartame is completely my own; the doctor has not confirmed it as aspartame-related."

"Restaurant Syndromes"

The subject of "restaurant syndromes" was mentioned in Chapter 25.

- A well-known example is the so-called "Chinese restaurant syndrome," usually representing a reaction to monosodium glutamate (MSG).
- The syndrome of sulfite sensitivity is recognized.
- Another is ciguatera toxicity after ingesting ciguatera-affected fish. It is characterized by vomiting, diarrhea, abdominal pain, joint and muscle pains, severe dizziness, numbness, and weakness.

Cartoonists and humorists have not overlooked this fertile topic.

- Michaelides noted in the March 1984 edition of Restaurant Hospitality that the presence of fake flowers on tables, and plastic plants hanging from the ceiling, should be a giveaway "of what to expect from the kitchen."

- The hostess at a Chinese establishment asked a couple whether they wished to be seated in the "MSG section" or the "non-MSG section."

Severe gastrointestinal reactions also have been precipitated by aspartame products. Nausea, vomiting, abdominal pain and diarrhea may be followed by headache, a rash, or even convulsions.

Accordingly, many aspartame reactors understandably harbor fear over being inadvertently served this synthetic sweetener in a dining establishment, regardless of reputation or assurances to the contrary. Some carry just-in-case plastic bags in their purse and car for the same reason that planes routinely provide airsickness containers.

Serving A "Diet Mickey"

Syndicated columnists receive their share of mail about reactions to aspartame-containing products. A letter from "Delaware Reader" to Dear Abby (The Palm Beach Post August 10, 1990) provided yet another perspective on this "restaurant syndrome."

> The writer was seated near a "plump" woman who ordered a regular cola at a fast-food restaurant. After this customer departed, the young waitress joked to Delaware Reader about having substituted a diet cola "because she was too fat, and would never know the difference anyway." She summed her reaction in one word: "reprehensible." This comment was reinforced by the fact that the writer predictably suffered severe headaches after drinking diet sodas.

The Slender Trap Syndrome

Serious disorders have accompanied the marked weight loss in some patients who resorted to "diet" aspartame products as a "culinary bypass" (see 24). Other possible components of this "slender trap syndrome" may include severe headache, loss of vision, intense depression, abnormal heart action (palpitations), and a host of complaints listed in the Overview.

Fellow Sufferers, Unite!

Some "aspartame victims" were both overwhelmed and gratified on discovering that others promptly recognized their syndrome. To a large extent, this has been the glue of aspartame consumer groups.

> A young woman suffered severe visual and neurologic complaints while using aspartame products. She had been diagnosed as having "probable multiple sclerosis." She wrote of her encounter with the founder of an aspartame group.
>
> "When I initially discussed the problem with _____ , she knew exactly what I was talking about. She had experienced much of the same problems. This was incredible to me because I had never imagined there would be someone who could match me word for word, symptom for symptom, about what I had been going through."

34

MOUSE OR MAN?
MAYBE THE RATS WERE RIGHT!

People who studied my previous writings on both aspartame reactions and other subjects have asked, "Does the FDA really equate human beings with rats and mice when testing the safety of new foods and additives?"

This man-versus-mouse question is not simply an academic matter. The Delaney Clause — incorporated in the 1958 Amended Food, Drug and Cosmetic Act, and the 1960 Amended Color Additive Act — specifically bans chemicals that cause cancer in animals from foods, drugs and cosmetics used by humans. Specifically, it states: "No additive shall be deemed safe if it is found to induce cancer when ingested by man or animal, or if it is found, after tests which are appropriate for the evaluation of the safety of food additives, to induce cancer in man or animal."

In spite of their shortcomings, mice experiments cannot be ignored. Indeed, they have played an important role in medical progress.

- A half century ago — specifically, the morning of May 25, 1940 — Howard Florey conducted the famous mouse protection experiment in eight white mice that was to establish penicillin as a "miracle drug."
- The recent availability of E5 murine monoclonal IgM antibody to endotoxin for the treatment of severe life-threatening infection by gram-negative bacteria represents an enormous advance in medicine. It is produced by the so-called heteromyeloma cell that results from combining a mouse myeloma (tumor) cell with a human myeloma cell.

This sobering chapter also serves as a preface to Chapter 44, "Who Killed the Sugar Plum Fairy?"

Pre-Licensing Testing

The studies dealing with possible toxicity, tumor formation, birth defects and other biological problems after aspartame administration were largely limited to experimental animals before this compound was licensed as a Generally Recognized as Safe (GRAS) additive in July 1981. I am not familiar with extensive tests on humans before such approval by the FDA Commissioner, notwithstanding the legitimate doubts and unanswered questions raised about aspartame's safety by several in-house FDA scientists and a distinguished Public Board of Inquiry (PBOI).

The Acceptable Daily Intake (ADI) of Aspartame: Another Myth

Fairy tales, such as the one in Chapter 44, can be dismissed as myths. On the other hand, our licensing and regulatory agencies must not be "myth-led."

Which brings me to the FDA's arbitrary increase of aspartame's ADI to 50 milligrams (mg)/kilograms (kg) body weight. (One kilogram is 2.2 pounds.) This questionable decision qualifies as an action that succeeds in "canceling reality."

- The ADI represents the projection of <u>animal</u> studies based on their <u>lifetime</u> intake. This was the testimony of Dr. Frank Young, former FDA Commissioner, before a committee of the U. S. Senate at a hearing on November 3, 1987 titled, "NUTRASWEET"—HEALTH AND SAFETY CONCERNS."
- Market research indicates that diabetics use about 11.4 mg aspartame/kg daily (<u>The Palm Beach Post</u> March 8, 1990, p. D-13).
- The vast majority of my patients with severe reactions attributed to use of aspartame products got into trouble when their daily aspartame intake ranged from 10 to 18.3 mg/kg. Once this threshold was exceeded, they <u>predictably</u> suffered itching, rashes, severe headache, mental confusion, depression, visual problems, et cetera. (Hundreds of

others experienced adverse effects after consuming
amounts far below these levels.)

Unfortunately, we haven't as yet been able to develop species of
mice or rats that can communicate symptoms. They can't even tell us
whether what they're being served actually tastes sweet!

When it comes to answering another question — viz., "Which species is the better metabolizer of phenylalanine?" (see 26) — rodents win
hands down by a 5-to-1 margin. This also applies to their ability to
handle methyl alcohol, another component of aspartame. Stated differently, rats aren't good predictors for adverse reactions by humans to
aspartame products.

The Matter of Saccharin

Okay, I admit to having a tendency to fixate upon "minor details" in
the evaluation of scientific data. One such item focuses upon gender
vulnerability in rodent studies concerning induced cancer.

This particular interest was sparked by a famous (or infamous) 1973
report of the Wisconsin Alumni Research Foundation. Urinary bladder
tumors were found in a few male rats given large amounts of saccharin.
There are two flaws. First, the validity of this rat model has been
challenged by experts. Second, other researchers haven't been able to
reproduce these findings.

So, you ask, "What's the big deal?" It became a real big deal when
bureaucrats pounced on this mouse-to-man controversy by invoking
the Delaney Amendment relative to use of saccharin. The FDA mandated that products containing it were to be labeled as potentially causing cancer in man! The producers of other sweeteners clearly were not
displeased.

Unfortunately, bureaucratic arrogance in this matter persists despite
firm statements that there is no convincing evidence saccharin in cus-
tomary amounts causes cancer in humans! These position papers have
been released by the American Cancer Society, the American Medical
Association, and the American Diabetes Association.

Recent studies suggest that the mechanism for development of bladder cancer in rats given saccharin involves the formation of toxic crystals in their bladders. On the other hand, humans lack the urine protein levels needed to form such crystals.

More "Minor Details"

Other "minor details" that were overlooked or ignored by the FDA prior to approving aspartame also generated considerable concern.

First, rats given aspartame developed a disproportionately high number of brain tumors. Spontaneous brain tumors are quite rare in laboratory rats, especially those younger than 400 days. Yet, the then-federal attorney and his successor did not prosecute two critical studies analyzed by a grand jury investigation of the drug company "for concealing material facts and making false statements in reports of animal studies" in a timely manner. As a result, the five-year statute of limitations expired.

> This lawyer quit as U.S. Attorney before these dates, and began working for a law firm representing the manufacturer! He is currently on the White House staff.

Second, the aspartame-induced brain tumors were found primarily in male rats. But aspartame and its diketopiperazine derivative apparently had been administered ONLY to female rodents in several of the carcinogenicity studies.

And be particularly careful of funny-looking humans with long white coats; they could give you cancer!

Cartoon 34-1

214

Cartoonists have raised an interesting possibility that is right on the money: rats may be more aware of such an inherent danger than Washington bureaucrats. Cartoon 34-1 conveys this message.

Humorists and cartoonists alluded to other aspects of this theme.

- One CEO admitted that his product had resulted in the death of mice. He qualified such an outcome, however, by duly noting that the creatures died content and happy.
- This issue could provide males with more ammunition in the event a "men's lib" counterrevolution were to gather momentum.

Have the Rat Studies Come Home to Roost?

I began to suspect the ultimate clinical boomerang of these ignored tumor observations in rats. The matter crystallized early in 1990 when reviewing United States cancer statistics. To my astonishment, <u>there was an unequivocal rise in the incidence rates of the more common primary malignant brain cancers since 1985...perhaps as early as 1984</u>. This could be shown for <u>all</u> races and <u>both</u> genders in the National Cancer Institute's 15-year SEER data, the most comprehensive analysis of cancer surveillance in the world.

Furthermore, there had been highly significant <u>annual</u> increases in the Estimated Annual Percent Change over the 1983-1987 period. (1987 is the last year for which complete data were available when this manuscript was finalized.)

Before you conclude that I'm paranoid, Dear Reader, please consider these facts:

- The foregoing experimental studies in rats were performed the 1970s. Even though they revealed a high incidence of brain tumors in rats, the 1958 Delaney Clause was not invoked. (Senator Howard Metzenbaum and investigative reporters detailed some of the highly suspicious political and corporate circumstances.)
- To my knowledge, no reports of additional studies

attempting to prove or disprove these findings for "the most tested additive in history" have been published since 1981.

- A Public Board of Inquiry expressed its unequivocal opinion that aspartame had <u>not</u> been shown to be safe as a food additive. It then recommended that the FDA delay approval pending the results of additional studies based on proper experimental designs.
- Aspartame products were licensed for human use as the "wet" form in beverages and for other products during July 1983. The cited increase of common primary brain cancers (astrocytomas, glioblastomas) began within one or two years. In my opinion, this increase cannot be attributed solely to better diagnostic methods since reasonably good brain scanning techniques had been widely available for at least one decade.

Additional evidence seems to validate the finding of brain tumors in rats that were given aspartame. <u>The 1982-1984 SEER data indicated a nearly threefold increase in the incidence of primary brain lymphoma</u> (a rarer subgroup) <u>among immunologically normal persons</u>. Specifically, the rate increased from 2.7 cases per ten million population in 1973-1977 to 7.5 cases per ten million in 1982-1984. Here are some relevant facts:

- Aspartame was licensed by the FDA for "dry" use in July 1981. Its "prodigious" consumption since then (see 20) is history.
- The age-adjusted rise was more striking among women — namely, from 4.9 per ten million in 1979-1981, to 8.9 per ten million in 1982-1984. (My data clearly indicate a 3-to-1 female preponderance among persons with reactions to aspartame products.)
- No other "confounding variables" as yet have been identified by investigators of this epidemiologic puzzle, including a number of neurotoxins in the environment.

Any reader who is miffed by this rodent linkage can find further details in my article titled, "Does Aspartame Cause Human Brain Cancer?" It was published in the Winter 1991 edition of the <u>Journal of Advancement in Medicine</u> (pp. 231-241).

35

AN UNLIKELY MEDICAL RAMBO

A short Epilogue in my earlier book ASPARTAME (NUTRASWEET*):IS IT SAFE? was titled, "The First — and Last — Word." One sentence therein invited reflection by readers: "I had no aspirations of becoming a majority-of-one physician counterpart of Rambo for 'aspartame victims.'"

If the basic public health problem weren't so serious, the idea of "Roberts As Doctor Rambo" could be fodder for standup comedians and cartoonists. In the real world, however, persons with adverse reactions to aspartame products increasingly cast me in this role against a billion-dollar industry. So did a number of consumer groups. I had to remind them that my motivation stemmed from concern as a practising physician, and not any vendetta against corporate America.

The Professional Vacuum

Three matters added fuel to this quasi David-and-Goliath scenario.

- First, the number of "aspartame victims" continued mounting.
- Second, the ranks of physicians who previously had expressed clinical interest in these matters dwindled as they envisioned high-noon confrontations with either the industry (see 11) or the media (see 29).
- Third, most of the initial investigators dissociated themselves from this controversy.

I sensed this evolving situation at a national press conference held in the Dirksen Senate Building during 1986 (see 23). No doubt remained when the Aspartame Consumer Safety Network asked me to address the subject during March 1990 (see 36).

The agenda of "Doctor Rambo" grew like Topsy. What began as a simple plea to other physicians about developing awareness of aspartame disease extended to the FDA. This pertained to (1) insistence upon proper precautionary labeling of aspartame products (e.g., an expiration date), and (2) the recommendation of a delay in licensing other "fake foods" until they had been adequately tested in humans.

Old Labels Never Die; They're Just Recycled

The number of complaints from infuriated consumers about improper labeling also kept escalating. One theme involved the alleged corporate dictum, "Ye Shall Not Waste!" A case in point was the use of old labels after producers added aspartame to beverages previously sweetened with saccharin. Those aspartame reactors who had carefully read all labels received a great surprise when experiencing "that terrible feeling" under these circumstances.

> A diabetic female developed severe headaches and confusion whenever she drank beverages containing aspartame. She later suffered an unexpected recurrence after ingesting only one-fourth of a can of diet soda. ("I felt my brain was swelling and too big for my cranium.") She then described the details of her investigation.
>
> "I called the bottling company and asked if they were using aspartame. They said yes! I told them the can said saccharin. They told me they were just using up the old cans, and then would change the contents as listed. I didn't get any satisfaction from them, so I called the Food and Drug Department in Oahu. I told them the story and they just said, 'Yes, that's what they do.' They were very uncooperative. I felt I was hitting my head against a brick wall. They didn't care a thing about me."

An Attitude Problem?

The chance viewing of a Rambo film on television evoked another chuckle relative to my purported "chemophobic" attitude (see 26). It was spawned by Rambo's classic reply about having "just a little problem with attitude" as he singlehandedly confronted some of the world's evils.

Forebodings About a Sciencegate Cassandra

In Greek mythology, Cassandra was the daughter of the King of Troy. Apollo bestowed the gift of prophecy upon her. In a subsequent fit of anger, however, this god decreed that no one should believe her prophecies.

I perceived an analogy over the apparent indifference of both the medical profession and the FDA to my warnings about the suspected public health consequences of aspartame reactions. It seemed that none of the publications, formal scientific addresses or media interviews by "Doctor Rambo" were being heeded in this example of Continuing Medical Ignorance (CMI).

As my data mounted, the question assumed more poignant overtones. I asked myself, "Will I become a modern Cassandra who cannot stem the potential heavy toll in human suffering before others either confirm or explain my concerns?" Foremost among these recommended priorities were the following:

- The need for repeating all earlier studies in animals and humans by corporate-neutral investigators who administered "real world" aspartame products (see 26)...with special emphasis upon brain cancer (see 34).
- The issue of accident-proneness among drivers and pilots due to confusion induced by these products.
- The possibility that aspartame-related memory loss might be a valid human experimental model for "early Alzheimer's disease" (see 19) — especially in view of the considerable aspartic acid found within the brain plaques of patients with this disorder.
- The apparent failure of the FDA to have learned a lesson from its premature approval of aspartame when it approved newer "fake" foods and additives that had not been sufficiently tested, if at all, in humans (see 43).

These "Sciencegate" inferences emphasized another need: the use of impartial "umpires" in any subsequent laboratory and clinical studies involving the foregoing challenges...even if it meant calling Sylvester Stallone in consultation. Sidney Harris nicely satirized this theme in a cartoon appearing in the March/April 1990 issue of <u>American Scientist</u>.

Cartoon 35-1

36

ROBERTS DOES DALLAS

There are times when one has to trust his or her instincts in traveling the uncharted seas of life, and "go with the flow." Such was the case when I decided to visit Dallas during 1990... specifically on March 6 and 7.

My trip chiefly reflected an expression of gratitude to Mary Nash Stoddard, President of the Aspartame Consumer Safety Network (Figure 27-1). She came to my attention after returning the nine-page questionnaire (see 22) in great detail, coupled with considerable information I hadn't seen. Her daughter also suffered convulsions from aspartame products.

Mary repeatedly requested that I visit Dallas to discuss my researches with several interested groups and talk show hosts. It was hard to refuse this spirited individual with such an appealing Southern accent, but I had to take a raincheck each time.

My indebtedness to Mary kept mounting as she relayed more data. Her organization went from a local project to one of national scope over these two years. I finally acquiesced and made the trip rather than suffering the alternative one — a guilt trip.

The number of radio and television shows that Mary managed to compress in two days attested to her PR ability. She also arranged a book-signing event and press conference at an extraordinary food market that refused to sell aspartame products out of concern for its customers. This session was recorded by NBC and ABC television affiliates. I realized that Mary's efforts had struck national gold when my niece from Wilmington, North Carolina, called after seeing the interview (see 21).

This sojourn to the Lone Star State also proved an emotionally rewarding one. I was warmly received by persons who had come to regard me as a combination of professional benefactor and "Doctor Rambo" (see 35).

I made another interesting observation while being chauffered.

Having heard reference to the X-rated movie <u>Debbie Does Dallas</u>, I turned a bit crimson on noticing the name of a major thoroughfare: Lover's Lane.

Mary Nash Stoddard

My review of Mary's records indicated that her recent life hadn't been an easy one. In addition to having suffering devastating reactions to aspartame products, her husband died from melanoma. This thrust the responsibility of rearing three children on a limited income upon her shoulders.

A vacuum was created when Shannon Roth of Ocala (Florida), founder of Aspartame Victims and Their Friends, halted all related activities during litigation...including her "hot line." Mary picked up the ball, created the Texas organization, and then transformed it into a national consumer network.

The gathering of clinical material and new scientific data, distributing them to interested parties, and educating pilots and pilot unions about potential hazards from the confusion and other adverse reactions to aspartame products represented a massive commitment. All the more so for <u>one</u> person working out of her home without any grant. But Mary handled these tasks with remarkable efficiency and professionalism.

My readers by now are probably aware of the fact that I am a fan of Ogden Nash. Since Mary's middle name was Nash, I had to resolve this detail. Were there any family ties with this poet? No.

Teaching the Teacher

Having been the invited guest on a number of radio and television shows (see 23), I labored under the delusion of acquiring some expertise in confronting the media. Recognizing the need for a few more "pointers," Mary took it upon herself to become a teacher — albeit in the restrained manner becoming a Southern lady. Some examples:

- "Try to direct an occasional answer for a question from the talk show host to the audience, such as, 'Your listeners might wish to know...'"
- "If someone asks an obviously hostile question, you

223

> might preface your answer with the phrase, 'But more importantly...'"

For good measure, Mary later threw in another pearl when she learned that ABC Television was planning to interview me on its <u>Home Show</u>.

> I was about to leave for a seminar in Waterville (Maine) when the producer called. She indicated that arrangements had been made to have a satellite dish brought to City Hall because ABC had no affiliate in the area. The interview was to be televised from the Council Chamber.
>
> Mary offered one surprising suggestion. "Sit next to an American flag!"
>
> When I arrived, the crew from Providence (Rhode Island) already had positioned my chair. The flag was a few feet away. I smiled as the banner was moved next to my left shoulder, thinking that Mary probably would be as pleased as if it had been the emblem of the Lone Star State.
>
> The proximity of Old Glory had an extraordinary, and unanticipated, gratifying effect upon me in this role as volunteer consumer advocate. Its impact was reinforced when I took the medical director of The NutraSweet Company to task for his biased corporate interest after he lambasted my concern about the frequency and significance of reactions to aspartame products.

Manning the Phones

En route to Channel 33, Mary mentioned that she had enlisted the assistance of her mother, Clara May, in handling the "hot line." This unusual request stemmed from anticipating many calls in response to our appearance on this major TV station. Furthermore, the Aspartame Consumer Safety Network had circulated the following release.

PANEL OF EXPERTS CONVENES IN DALLAS MARCH 7TH FOR ANNOUNCEMENT OF NEW RESEARCH RESULTS REGARDING ASPARTAME (NUTRASWEET/EQUAL™)

Representatives of a National Consumers Coalition, representing over 10,000 concerned consumers, are meeting with a noted aspartame researcher to announce their joint findings. The Aspartame Consumer Safety Network of Dallas & Washington, DC is joined by the Community Nutrition Institute of Washington, DC and Aspartame Victims and Their Friends based in Atlanta, to present statements of support for researcher/clinician Dr. H. J. Roberts of The Palm Beach Institute for Medical Research in West Palm Beach, FL. Author of a newly-published book, Aspartame (NutraSweet*):Is It Safe?, Roberts will review the results in over six hundred aspartame reactors and patients.At the same time, the consumers coalition congratulates the $65 million dollar food retail corporation, Whole Foods Markets, on its official announcement today of the decision to ban aspartame products from Whole Food Market shelves. (Their stores are located in Texas, Louisiana and California.)

Representatives of Pilots Concerned About Aspartame & Flying will be available to talk about their own reactions, and what the FAA is or is not doing to address their vital questions about safety for pilots and passengers. Multiple warnings to pilots have been published in their worldwide publications.

These announcements will be made public at a press conference, Wednesday, March 7, 1990 at 1:00 PM at Whole Foods Market, #60 Dal-Rich Village at Coit & Beltline, Richardson, Texas.

More on Clara May

After mentioning Clara May's special call to duty, Mary described her as a "tiny, frail and very sweet" lady who had served as a church secretary for many years before retiring. She then casually mentioned an anecdote I found both pertinent and <u>very</u> funny.

> It concerned the machine at the church that dispensed cans of soda pop. Aware of her daughter's and granddaughter's severe reactions to "diet" drinks, Clara May was not about to become an accessory to the repetition of any similar problem involving other members of <u>her</u> church. So, arms outstretched, she blocked the passage when a husky fellow came to replenish the supply of these beverages.

I did not reveal my immediate reaction to Mary on hearing this confrontation. Now it can be told: "That sounds like a <u>great</u> scenario for a cartoon in my new book."

Not Quite Paranoid

The NutraSweet Company monitored the activities of "Corporate Enemy #1" (see 11), especially my radio and television interviews. I was so informed by talk show hosts in West Palm Beach, Miami, and now Dallas.

These hosts had factual reasons for such awareness. For example, the Company would insist upon "equal time." Some producers acquiesced; others wouldn't budge...pointing out that this wasn't a political campaign.

> One Dallas host actually told his audience, "I guess our guest can't be accused of paranoia about being followed by the company because he really is being followed."

Mary and I were surprised by another encounter on March 6 at TV Channel 33 (KDAF). The NutraSweet Company had flown in a PR person from Chicago, and a Ph.D. nutritionist from San Francisco. The latter was interviewed after we left the set.

Several interesting things happened the next day. The first took place during my hour-long interview on KLIF. The nutritionist called in, identified herself, and then challenged all my views. Here are some highlights of her remarks, for which Kevin administered the third degree.

- She admitted, albeit very reluctantly, that she was not a medical doctor.
- She averred to having come from San Francisco at her own expense. "Why?" The answer: "It's a free country, isn't it?"
- Although pregnant, this nutritionist stated that she had no misgivings about consuming aspartame products while in this delicate state — a view diametrically opposite to mine.

Another corporate encounter occurred at the Whole Foods Market in Richardson. Just prior to the scheduled press conference, mentioned earlier, Mary noticed two persons "in disguise" near the frozen food section. The man was wearing a trench coat. It then dawned upon her that this was the same twosome she had met at Channel 33!

The Courage of Corporate Conviction

The Whole Foods Market scene proved extraordinary on several accounts.

Mary Stoddard sent me one of the fact sheets that this food chain had created for customers. It was titled, "ASPARTAME — HOW SWEET IT ISN'T." The flyer reviewed the basic evidence that caused this firm to discontinue selling aspartame products. It also emphasized the FDA's acceptance of "flawed studies." Yours Truly was duly impressed.

My reaction on entering this market could be described as a combination of amazement and envy. I was amazed at its cleanliness, the beautiful displays of fruits, vegetables and other merchandise, and the obvious intelligence and sophistication of those shoppers present. The envy reflected lack of awareness of any comparable store in my own community.

> A déjà vu occurred several years later while reading a report about a survey in Working Woman Magazine. It noted surprising correlations between one's education and the time taken to scan shelves of markets. Women with postgraduate educations spent the most time grocery shopping...averaging 8.4 hours a week. It dropped to 6.7 hours for those with only a high school education. An opposite trend, however, was found among men — viz., only 3.8 hours for those having graduate degrees, compared to 5.4 hours for males with undergraduate degrees.

I then met the manager and other representatives of this market. They were bright and interested persons who seemed highly committed to their calling. None expressed concern over the economic loss incurred by not selling aspartame products. That's commitment!

228

The Conference

The media interview was held around a table. A representative of Whole Foods Market, a pilot, and a dental technician who had suffered aspartame-related convulsions joined Mary and me. Each member of the panel candidly expressed his or her views.

I added a new wrinkle to the discussion when a proverbial "light-went-on" thought occurred while being asked about the options of consumers confronted with displays limited to aspartame soft drinks. I answered, "Just say 'No!'" (It also could be spelled "Just say 'Know.'") This was the first time I uttered the now-classic battle cry used in "the war against drugs."

One of the television interviewers then asked Mary about her group's recommendations. She reinforced my emphasis on the foregoing reply by stating, "Reclassify aspartame as a drug, which of course it is." (p. 176)

A Sweet Revelation

A bystander at the press conference could have been a model for the prototype of a lovely grandmother. I also noted her extreme attentiveness during the session.

She approached me as I was about to depart. "What are your thoughts about Stevia, Dr. Roberts?"

If I don't know the answer to a question, especially when posed by an intelligent patient, I generally admit it...pronto. On this occasion, I replied, "I really don't know what you're referring to, and would very much like to have you tell me."

Teacher thereupon became student. The woman pulled out several articles about a "sweet herb" from her purse. It bore the scientific name Stevia rebaudiona. She explained that it had been used by natives of Paraguay for more than a century to sweeten drinks and foods. The dried leaves and twigs of this plant not only were sold for this purpose in South American markets and pharmacies, but also for use by diabetics.

The lecture continued. A Stevia extract had been marketed as a nonsynthetic natural sweetener in this country by a Dr. Chen. The FDA then withdrew its approval as a food additive during July 1984. It did allow use of this extract, however, as "a cosmetic for external use"...particularly facial massage. Since the basic glycoside is 300 times sweeter than table sugar, stable in heat (contrasting with aspartame in this regard), and has an indefinite shelf life, I wouldn't be surprised if some advocates were satisfying their sweet tooth with this "natural facial masque."

Over the ensuing years, the low key entrepreneurship for Stevia leaves and extract impressed me It had been stimulated by (1) its successful use as a sweetener in Japan as well as South America, and (2) reports of dramatic responses in patients with hypoglycemia, diabetes, nonspecific lethargy and other disorders.

> A knowledgeable fellow in this business called me while en route to Paraguay after reading several of my publications on aspartame reactions. His small Stevia operation had been halted by the FDA, even though no toxic effects were known to him or to this agency. A retired federal attorney later confided that a barrage of corporate attorneys representing a manufacturer of aspartame had forced such arbitrary action, which he considered groundless.

Grilling by a Hostile Host

The host on television Channel 33 (KDAF) admitted his partiality to aspartame products right up front. He emphasized the delight of drinking three cans of diet sodas daily. This prepared us for the not-so-subtle hostility that surfaced during our live interview. For instance, he asked about Mary's income as a "paid consultant." (Answer — none.)

An interesting development took place as this host then prepared to interview the cited Ph.D.-nutritionist who represented The NutraSweet Company...alone. To orient himself, he donned glasses before reading "fact sheets" affirming the alleged safety of aspartame products. I had not seen him wearing glasses during several preceding interviews. This caused me to wonder about a causal relationship between such need for visual assistance (see Overview), especially in front of television cameras, and his confessed obsession with aspartame beverages.

Lightning Strikes Twice

One of the "aspartame victims" I met in Dallas was Larry Taylor. I had known of his problems from various sources. They included the completed questionnaire survey he sent me, and his testimony before the U. S. Senate hearing during November 1987.

This 35-year-old nurse-anesthetist had suffered severe headaches, memory impairment, visual difficulties, and three subsequent convulsions while drinking 4-6 diet colas daily. At the Senate hearing, he related the following remarkable reply by a neurologist to his suggestion that aspartame products might be the cause: "Wouldn't it be a shame if all that is wrong with you is NutraSweet?"

Larry recounted the ensuing difficulty getting another job in his profession. He was finally hired when a senior anesthesiologist encountered another colleague who had suffered similar severe reactions to aspartame products. He told Larry, "Now, I believe you! You can work for me."

Larry considered suing the manufacturer. The statute of limitations expired, however, when he couldn't find a law firm willing to accept the case. One prominent Dallas attorney did indicate a willingness to do so, but his partners insisted on receiving $200,000 up front!

Needless to say, Larry exercised extreme care to avoid any product that might contain aspartame. For example, he studiously avoided all "diet" salad dressings in the hospital cafeteria. But the one time he lowered his guard proved catastrophic.

> Larry developed a cold. He could find no reference to aspartame on the label of a famous-brand "cold plus" concoction with a "lemony flavor" that contained aspirin, a decongestant, and an antihistamine. His face and mouth became enormously swollen within minutes after ingesting it. When his wife couldn't recognize his features, she wouldn't let their children look at him. Larry was taken to an emergency department, where he fortunately responded to an injection of epinephrine (adrenaline). The "lemony flavor" was found to contain aspartame.

I wouldn't be surprised to find a famous slogan placed in all strategic areas of Larry's home and car. It's the one that reads, "NEVER AGAIN!"

Dallas Revisited

I was pleased to be invited again to the hospitable Dallas area on several occasions.

I participated in a symposium on the safety of aspartame products sponsored by the University of North Texas during November 1991. Various anecdotes about this seminar and its ramifications appear in Chapters 6, 11, 14, 17, 24 and 30.

On March 28, 1992, I served as headline speaker at the 39th Annual National Convention of Natural Food Associates. The preliminary announcement referred to me as "the nation's most foremost medical authority on aspartame disease." I took pains to indicate that others had worded this brochure.

37

"REGRETS" VS. "NO REGRETS": REFLECTIONS ON WRONG-WAY CORRIGAN

This chapter on the subject of regrets, or the lack thereof, is "open-ended" because it remains to be determined which party ultimately will be expressing them.

A Preamble On Regrets

Yours Truly doesn't balk at extending regrets when they are warranted. After all, no one's perfect.

- I routinely apologize to patients who have been kept waiting 30 minutes or longer in my office. This even applies when there is a logical explanation for an unforeseeable delay — e.g., an intervening emergency, or an urgent phone call from the hospital or a colleague.
- I have a standing offer to retract any observation or conclusion in my publications if a critic (1) has done his or her homework, and (2) can demonstrate where my data or reasoning are wrong. Furthermore, this action will be accompanied by an expression of true appreciation to such a "teacher."
- A cartoon showing strange equipment, described as a combined word processor and food processor, contained a comparable sentiment. It had been devised for someone "who has to eat his words."

But there's another side to this equation. It was nicely phrased by Demosthenes: "I do not purchase regret at such a price."

"The CAA Regrets ..."

Returning to the subject at hand, a major basis for my concern over adverse reactions to aspartame products involves vehicle and aircraft safety. I offer my 1,000-page text, THE CAUSES, ECOLOGY AND PREVENTION OF TRAFFIC ACCIDENTS (Charles C Thomas, 1971), as Exhibit #1 for evidence of my interest in this subject.

Exhibit #2 consists of data indicating that aspartame-related problems can contribute to accident proneness by drivers and pilots. The most notable are confusion, memory loss, severe drowsiness, agitation, tremors, "blacking out" and convulsions. These observations appear in my medical publications and in correspondence published by General Aviation News (August 28, 1989).

Given this background, the reader can understand my dismay on reading the following statement released by the CAA during 1989.

> "To attribute such symptoms to Aspartame or any food recently ingested could be a serious error. As the item in GASIL [General Aviation Safety Information Leaflet] made clear, there had been no scientific evaluation to substantiate the cases reported in the USA. The CAA pointed out that colas contain caffeine which could, if taken in excess, cause similar symptoms. *The CAA regrets any misunderstanding that may have arisen."* (Italics supplied)

These "regrets" were influenced in large part by one "scientific" study presented by Belger et al at the annual meeting of the Aerospace Medical Association in May 1989. They gave 12 pilots pure aspartame, and then administered a cognitive test battery. The study concluded that the "results do not appear to support the concerns expressed in anecdotal testimony regarding cognitive performance." The abstract, however, did not indicate whether or not corporate sponsorship existed or if "real world" aspartame products also had been studied.

Non-Reciprocated Regrets

My response to the foregoing conclusion could have been predicted. I had challenged similar "negative" conclusions in other studies on the grounds that the protocols were flawed because <u>persons don't consume pure aspartame</u>. Rather, they ingest commercial products wherein the aspartame molecule could be profoundly altered by heat, prolonged storage, and other factors (see 17 and 26).

I made a conscious effort over the next week to count up to 100 whenever I pondered these CAA "regrets." Was it worth the time and energy to express "righteous indignation" to yet <u>another</u> bureaucracy? Conscience made the final decision.

I stated my views to the Civil Aviation Authority and the Federal Aviation Administration, as well as <u>General Aviation News</u> and <u>Plane & Pilot</u>. The latter ended

> "Until such time as adequate 'real world' studies have been performed by corporate-neutral investigators, I cannot in good conscience offer my own 'regrets.' After all, <u>I</u> may be one of the passengers at risk."

I was sorely tempted to add a "maxim" written three centuries earlier by Francois, Duc de la Rouchefoucauld. It seemed appropriate in the case of a governmental agency that had gone so far out on a pro-industry limb. But I decided against doing so for reasons of brevity.

> Here is Duc's assertion. "Our repentance is not so much regret for the ill we have done as fear of the ill that may happen to us in consequence."

Reflections on Wrong-Way Corrigan

The term "wrong-way Corrigan," mentioned in the title, was spawned overnight when a solo pilot by this name took off from New York in July 1938 and landed in Ireland rather than California.

This event occurred long before anyone had conceptualized the aspartame molecule. It nevertheless represents a possible, albeit remote, consequence of aspartame-induced confusion in pilots.

Covert Mail and Calls

I continued to receive mail from pilots about symptoms they attributed to aspartame products, along with their accounts of "near misses." A majority described "close encounters of the aspartame kind" caused by confusion and memory loss. Most elected to remain "anonymous" for fear of jeopardizing their licenses.

This theme is illustrated by a letter from one pilot who wrote of "immediate and severe effects upon my consciousness and vision" after drinking diet beverages.

> "I described these symptoms and circumstances to my doctor. He scratched his head, ran tests, but never seriously listened to my linking pop — at the time I was not suspicious of aspartame — with such effects. He was just an average pill-dispensing GP, reasonably competent, but not ready to distrust, let alone blame, an FDA approved sweetener. But he kept notes and must have figured me a health risk, since an insurance company refused issuing a life policy. When I asked why, they advised that their denial was based on my GP's report. I took him off my Christmas card list immediately. Eventually I quit diet drinks and have not had an incident since...While I agree that the DOT [Department of Transportation] should know, I feel that the poisoner (aspartame) should be grounded, not the poisonee (pilot). However, that would require the DOT to be more creative than my doctor in its investigation and analysis. I am not holding my breath. Perhaps anonymous writing is the answer."

More on Expressing Righteous Indignation

Unbeknownst to me at the time, Mary Stoddard of the Aspartame Consumer Safety Network (see 36) also had been writing to the editors of aviation magazines and pilot unions about this potential problem.

We subsequently exchanged views.

The following letter appeared in <u>The Palm Beach Post</u> (October 14, 1989, p. E-2) after it had published a report on the magnitude of serious plane accidents.

"Several recent plane accidents underscore the need for further inquiry into a heretofore-neglected cause of pilot and driver error: confusion and aberrant behavior caused by products containing aspartame (NutraSweet®). For example, did the copilot who inadvertently hit the disengage button before the recent USAir jet accident, and then acted 'irrationally,' ingest an aspartame 'diet' soda or coffee sweetened with an aspartame tabletop sweetener?...

"I have repeatedly pointed out these and related problems in scientific articles and addresses over the past 3 years. They are based on personal observations and a nationwide (non-granted) study. My report on 157 persons with aspartame-induced confusion and memory loss was published last month. The subjects included trained pilots who developed these and other neurologic or psychiatric features—including convulsions and visual problems. (These health-conscious professionals used such products in an attempt to avoid sugar.) I commented on these findings and their implications in the August 28 issue of <u>General Aviation News</u>. In my opinion, there is more justification for labeling aspartame products, 'WARNING: Use Caution If Driving a Vehicle or Flying a Plane,' than the disproved inference about saccharin causing cancer."

Intra-Professional Warnings

The word is getting around…finally. Flying Safety (May 1992, p.21) warned ALL pilots who consume aspartame products that they may be subject to memory loss, dizziness during instrument flight, gradual loss of vision, flicker vertigo, and flicker-induced convulsions.

Misconception About Conception

Humor occasionally punctuated my correspondence from pilots. Describing his experience with blurred vision and confusion after drinking two cups of aspartame-containing hot chocolate, one pilot wrote

> "As an interesting side light, a week earlier, one of the office secretaries had just gone on a diet and switched from a regular cola to a diet cola. While shopping, she had to take the shopping cart away from her husband to keep from falling down, and gave him the keys to the car because she felt unable to drive. When I related my reactions, she said that she had felt exactly the same. The editor of Pacific Flyer will appreciate her husband's reaction...'What's the matter with you, pregnant?'"

Straight From the Captain's Mouth

The following encounter of Barbara Mullarkey (see 23) deserves retelling.

Barbara was returning from the two-day symposium on aspartame safety held at the University of North Texas during November 1991. She happened to mention the seminar in a discussion with the captain of her plane. He replied, "Oh, we know about that! It causes spikes in the EKG."

This tidbit troubled Barbara all the way to Chicago. She kept asking herself, "Did he mean EKG for electrocardiogram or EEG for electroencephalogram?"

The captain greeted her as she was about to leave. Barbara thereupon asked him the question. He replied, "I did mean EKG." "But how do you know this?" Answer: "It happened to me! The medical examiner told me to avoid diet drinks before taking my physicals."

38

FROM THE MOUTH AND PEN AND MIND OF BABES

Let me again emphasize the extraordinary potential for adverse reactions to aspartame products among infants and children. They include convulsions, headaches, rashes, gastrointestinal conditions, impairment of intellectual performance, and aberrant emotion or behavior. Unfortunately, these associations are not generally recognized.

As far as I am concerned, it is never too soon to avoid exposing the young to aspartame products! In one case, for example, a suckling infant developed seizures in the arms of its mother as she drank an aspartame cola while breastfeeding. (Such children also could be correctly referred to as "coke babies.")

> A girl in Bill Keane's cartoon The Family Circus uttered a pertinent slip of the tongue when she blessed "this food that has been repaired for us" (The Palm Beach Post July 30, 1990).

This physician really doesn't want to give the impression of being a congenital "Dr. Kill-Joy." After all, my six children and four (so far) grandchildren have enjoyed an occasional treat of sweets. The problem, however, resides in the ongoing introduction of new aspartame products that have special appeal for them, such as sugar-free chocolate milk.

Kids as Diagnosticians
Some tots merit a Roberts Gold Star for realizing that their problems were related to the use of these products. In some instances, such awareness occurred long before their parents and doctors had even a conscious inkling about this connection.

An 8-year-old girl suffered terrible headaches every day, for which she repeatedly received aspirin. Her mother arranged for further medical consultation. At that point, the child pointed out that her headaches <u>predictably</u> occurred <u>ten minutes</u> after she chewed gum containing aspartame. As soon as it was avoided, the headaches ceased!

A Son's Emotional Reaction

The first-hand account of Kelly Allen's affliction with seizures due to "my coffee," which she titled "To Hell and Back on Aspartame," appears in Chapter 45. This young mother also enclosed a "cartoon" on which she typed this explanation: "Eric, 9, spent 5 hours at a creative writing workshop at school. This is what he wrote about. Eric was in the car in '87 when I drove off the cliff."

Cartoon 38-1

Mark Twain Reflects

The current widespread use of aspartame products by children warrants repetition of a true story involving Mark Twain.

> Copies of <u>The Adventures of Tom Sawyer</u> and <u>The Adventures of Huckleberry Finn</u> were removed from shelves of the children's room at the Brooklyn Public Library. The reason: Twain stated that he had written these books exclusively for adults, and therefore was distressed to learn that boys and girls were allowed to read them. He observed, "The mind that becomes soiled in youth can never again be washed clean."

The term "soiled" also might encompass the adverse changes in brain physiology caused by aspartame products to which the young are uniquely vulnerable.

School Research Projects

I cannot leave this chapter without mentioning a truly joyous dividend of my publications on the adverse effects of aspartame. It came in the form of letters from bright junior high school and high school students who used their imagination to expand upon my observations for "science projects." Here is one example involving flatworms.

> Dear Dr. Roberts:
> I am a twelfth-grader at Cocoa High School in Cocoa, FL. I am currently conducting an experiment of the effect of aspartame on the conditioned reflex of planaria. So far, I have performed two weeks of conditioning trials on the planaria. The planaria that received aspartame in their water did not learn the conditioned reflex as well as the planaria in the control group, which received no aspartame. I plan to run more trials on the planaria before the state science fair.
> I have read your book on aspartame. I found

the information in it extremely helpful. Any other information you can provide me with would be very much appreciated.

39

STRANGE PRESCRIPTIONS

I generally carry several prescription blanks in my shirt pocket. They come in handy when patients require new medication after being discharged from the hospital.

Yours Truly also utilized such "spare paper" for making notes during conferences, and recording anecdotes about reactions to aspartame products in nonmedical settings. A number appear in other chapters. Here's another.

> A prominent businessman asked me at a social function, "Doctor Roberts, does aspartame cause dry eyes?" Since the thought had never occurred to me, I reflexively jotted down his question. When I so queried aspartame reactors thereafter, several promptly replied in the affirmative.

Another lesson about the human condition motivated such behavior. Experience has taught me to distrust the notion that I "could not possibly forget" some important observation or story...only to have it fall right through the cracks of my memory.

A cartoon embodies the value of this "reserve scratch pad." It shows a mechanic telling the owner of a car, "You never know what you're going to find once you really start looking."

"Scribblers Anonymous"

I don't know if there's an organization called "Scribblers Anonymous." If it exists, an application for membership will be appreciated.

In addition to the foregoing "strange prescriptions," I resorted to scribbling insights on other forms of improvised "note paper." I had utilized this tactic nearly two decades ago during the initial top-secret phase of a literary project, My Wife, The Politician. It involved the covert recording of many delightful, but unpredictable, anecdotes about

Carol's activities. This often necessitated the use of blank checks, napkins, discarded envelopes, empty spaces in or around magazine ads, and the back side of laundry lists from hotels that didn't supply writing paper. I derived some consolation from the knowledge that famous authors, poets and even composers had used comparable "scratch paper."

- Francis Scott Key began writing verses titled <u>Defense of Fort McHenry</u> on the back of a used envelope. This work was later retitled <u>The Star Spangled Banner</u>.
- Carl Perkins wrote his original version of <u>Blue Suede Shoes</u> on a brown paper bag previously containing potatoes. Elvis Presley subsequently popularized it.
- Camille Saint-Saens noted a local African love song on his shirt cuff, which became incorporated in <u>Africa</u>.
- Abraham Lincoln allegedly composed or modified <u>The Gettysburg Address</u> on scrap paper while riding in a railroad car.
- Diane Warren wrote or co-wrote 18 "Top Ten" songs since the early 1980s. Describing how titles and ideas "just pop into my head," she would write them on the palm of her hand...or even Kotex (<u>Parade</u> Magazine, July 22, 1991). When <u>really</u> desperate, she would call home and sing to her answering machine lest the ephemeral inspiration evaporate into eternity.

A Non-Prescription Contraceptive

I do not practice obstetrics or gynecology. Few women therefore consult me about birth control pills and contraceptive measures. Even so, I was intrigued by a study published in <u>The New England Journal of Medicine</u> (Volume 313, 1985, p. 1351). It concerned the spermicidal effects of four cola products made by the same manufacturer.

This investigation involved checking the mobility of sperm (obtained from the semen of a healthy fertile donor) after exposure for one minute to these beverages. They were opened at room temperature. The averaged results of four measurements, compared to a control (saline solution), indicated marked reduction of sperm mobility by each soda. The aspartame product proved to be the strongest spermicide. Here are the results:

Sperm Mobility at 1 Minute
(expressed as % of control)

"Classic" cola soda 8.5
"New" cola soda 41.6
"New" caffeine-free cola soda 16.6
Aspartame cola soda 0

Surprisingly, I have not come across any "field study" reporting the contraceptive value of these products. Some entrepreneur surely must have considered this "scientific" answer to overpopulation, particularly in "third world" countries now being bombarded with ads advocating the virtues of aspartame sodas via the usual route of consumption. The absence of such articles suggested the two F's as an explanation: "futility" and "fertility."

Unfortunately, such a report could have inherent value for racists and rumormongers.

Posters kept appearing in New York City ghetto apartment buildings alleging that certain sodas were being produced by the Ku Klux Klan "with an ingredient to sterilize black men" (The Palm Beach Post July 7, 1991, p. A-9). Furthermore, their prices purportedly were kept way below those of well-known brands for this reason. For its part, the KKK adamantly denied any involvement in bottling monkey business.

40

ANSWERED PRAYERS

A repeated theme in scores of letters received from aspartame reactors focused on an "answered prayer."

There's at least one thing that can be said for prayer: it might do some good, and is unlikely to do any harm.

> As one who engages in formal daily prayer, I have some conviction about this uniquely human enterprise. Many patients expressed a higher comfort level after learning that their physician prays for his patients. Perhaps there's some connection with a slogan used by the Hebrew National Company, a manufacturer of kosher products: "We answer to a higher authority."

The Revelation

A 46-year-old woman wrote about her severe neurologic problems. She had undergone extensive studies for difficulty in walking ("I felt like I was on a pair of stilts"), numbness, loss of bladder control, severe headaches, depression, and "white spots in my eyes making seeing difficult." She added

> "Knowing I was facing all the medical testing, I asked our church membership for prayers. Within two weeks of that request on October 12, 1989, I had the strangest feeling — get off the NutraSweet. I did so. Within five days the fog lifted from my head and there was a lessening of the tightness across my hips and lower back...In this is my hope and prayers for a full recovery."

246

Prayers of Thanks

A 45-year-old woman had become incapacitated while consuming considerable amounts of aspartame products. Her symptoms included intense headaches, memory loss, confusion, depression, personality changes, recent hypertension, joint pains, itching, loss of hair, a gain of 75 pounds (!) (see 24), and intense thirst. She wrote

> "If my friends and family hadn't loved and cared for me, I hate to think what would have happened. My husband led me around like a child. I couldn't converse with others. I had problems finding the bathroom without help...

> "I then read an article about aspartame and stopped using it, and began improving. My headaches diminished and stopped after a few days. My aches and pains took a few weeks to improve. My memory took months...I worry if it will ever be the same. I am thanking God every day for having my mind back.

> "My sister's headaches, memory loss and vision have improved after stopping use of aspartame."

Reverend Pat Robertson

Reverend Pat Robertson (see 14) reinforced this attitude on The 700 Club by preaching what he practised: "Avoid aspartame products!" This was based on his own experiences. In the present context this theme could translate to "The Lord helps those who help themselves by following good advice."

The following "testimonial" was written on the last page of my questionnaire by a 41-year-old woman who had viewed one of these CBN programs. Her seizures began three months after starting to use aspartame products, and disappeared when she avoided them.

"Multiple tests were done and no reasons for my seizures were revealed. Tegretol® was prescribed. The Lord revealed to me I didn't need the medication and I trust Him!! At the same time, there was a special report on CBN about the effects of aspartame. It was most enlightening."

Praise the Lord...and Pass the Ammunition

A number of persons who recovered from severe reactions to aspartame products not only expressed gratitude for this "divine revelation," but also waged their own proselytizing crusades. The extraordinary recovery of Jan Smith from aspartame-linked hyperthyroidism was recounted in Chapter 6. Ann Topper, a friend and "victim," subsequently wrote her:

"I'm so proud of you for taking the initiative to make yourself well contrary to the well-meaning but incompetent medical establishment. Remember, after all, they're just "practicing" medicine. LISTEN TO YOUR BODY! Congratulations on your return to health and good luck with your endeavors in promoting public awareness of this aspartame problem. I tell everyone I see!"

41

FANTASY ENCOUNTERS WITH STARS AND PROMOS

An analysis of TV ads extolling the virtues of low-calorie aspartame beverages and foods should be required study for Marketing 101. Famous movie stars, television personalities and other celebrities have promoted these products, particularly "diet" beverages.

- Raquel Welch was mentioned in Chapter 18.
- Vanna White, Billy Crystal, Joe Montana, Ray Charles and Paula Abdul have touted diet cola drinks.
- A blitz of TV ads for a popular aspartame tabletop sweetener featured Cher.
- Even Roger Rabbit, the animated character, left the brier patch to make a similar pitch.

The hype promoters learned that their efforts can be overdone. A "Pop Quiz" evolved when it became difficult to follow all the celebrities being featured in cola commercials. The Wall Street Journal (January 24, 1990, p.B-1) reported that one cola producer used 27 celebrities and 31 NFL football players in its 1989 commercials!

Such stellar associations from "myth-led" television commercials could have a lasting negative impact on children (see 38). It should not come as a surprise that many learn their NBC's and CBS's before their ABC's. The younger generation hopefully will not confuse "light" or "lite" sodas with "light" beer (see 4). The same could be said for "light" whiskey since Seagram's introduced Mount Royal Light.

Truth In Endorsing

Crazy People, a comedy film starring Dudley Moore and Daryl Hannah, focused on truth in endorsing. I was jolted by the accuracy of this paraphrased remark by an advertising mogul: "You can't level with the public because this is America!" In point of fact, it's becoming harder to believe endorsers when "lite" can refer to alcoholic beverages (see above), and the inference is made with a straight face that aspartame is a "natural" substance (see 26).

But actor beware! Legal hitches can embroil celebrities who accept payment to endorse products or services that are later alleged to have caused injury or economic loss.

> The U. S. Bankruptcy Court of Chicago ruled that such persons could be held liable for misrepresentation if they failed to take reasonable steps to ascertain the truth of statements they uttered in advertisements (The Palm Beach Post February 8, 1990, p. B-19). This court rejected the excuse that these "household names" were "merely reading a script."

A Courtroom Scenario?

Another scenario triggered my imagination. What if I agreed to serve as an expert witness for a plaintiff who claimed medical harm from aspartame products that had been promoted by celebrities? Would I become tongue-tied if some glamorous movie star gazed intently at me while expressing such an opinion "with reasonable medical certainty"? Would I be intimidated by the dagger-like glances of a box office Super Macho star skilled in the martial arts?

Close Encounters of the Aspartame Kind

Speaking of movies, Mary Nash Stoddard, President of the Aspartame Consumer Safety Network (see 36), provided a related story antedating her involvement as a consumer advocate. She had consecutively consulted a family physician, an internist, a neurologist, an ear specialist, a gynecologist, and an audiologist for her as-yet-unrecognized multiple reactions to aspartame products. Mary summarized the results: "Nothing concrete was found."

When Mary later discovered that many other persons had similar afflictions attributable to aspartame products, she wrote

> "Remember the movie, Close Encounters of the Third Kind by Steven Speilberg, where Richard

Dreyfuss is excessively molding mounds of mashed potatoes and garbage into the shape of Devil's Monument, and everyone (including his own family) thinks he is crazy? Even <u>he</u> thought so until he went to Devil's Monument and saw other people to whom this strange thought had also occurred. Then they and the entire world knew that they were not crazy. They got together and proved it to everyone!"

Subliminal Messages

Sexual attraction clearly motivates the choice of glamorous actresses and handsome movie stars for many sugar-free promos. But there are limits involving "community standards" in related matters, such as the design of pop cans. For example, some infuriated parents saw the word "sex" when the computerized pattern of Pepsico Inc.'s "cool can" was stacked in a certain way. It had been released as a collector's item in the summer of 1990.

More on "Uh-Huh"

The "Uh-Huh" sung by Diet Pepsi frontman Ray Charles became incorporated into T-shirts, sweatshirts, and even silk neckties. Its popularity also stimulated research concerning the origins of this term...originally spelled "Uh-hu." A born-in-America expression for "yes," it has been more formally referred to as a "grunt of affirmation" used by the Tolkapaya-Yavapai of Arizona and other native Americans.

42

THE FDA – FRIEND OR FOE?

We need a competent and effective Food and Drug Administration (FDA) to ensure the safety of drugs, foods, additives and cosmetics. Since laughter is considered good medicine (see Preface), another sober consideration has been raised by some critics of this agency: "Will humor be next on its regulation list?"

The powers of the FDA stem from previous gross abuses and tragedies. Examples include the toxicity of sulfanilamide in the 1930s, and severe birth defects associated with the use of thalidomide during pregnancy. In the latter instance, the good fortune of the United States did not represent purposeful action by the FDA, but rather "timely inaction" by a member of its staff. (Some cynics aver that FDA also stands for Foot Dragging Administration.)

Let me make the basic point here, Dear Reader. You probably would not be reading this book if there had been "timely inaction" before approving the artificial sweetener aspartame.

I know that the FDA gets lots of flak from all sides as a target for consumers, consumer advocates, manufacturers, and the medical profession. Admittedly, some of the charges about its policies, decisions and surveillance have been unfair or unfounded. Such criticism appears justified, however, relative to the perception that the FDA has favored Big Business in its approval and continued sanctioning of aspartame as a sweetener (see 15). The consuming public needs to be assured that the FDA is on its side when confronted with a phenomenon some have captioned "America's Sweetener Armageddon." The appeal of ads for aspartame-sweetened products, now a gold standard for the PR industry, raises doubts about the FDA's effectiveness in confronting powerful business interests.

Bureaucratic Revelations

Many persons who became concerned over the safety of aspartame products got their first unflattering revelations about the FDA and other regulatory agencies as a result of such interest. It became abundantly clear to them that these agencies — numbering about 55 headquartered in Washington, D.C. — serve to protect the competitive market place, not the consumer.

The joking assertion that Washington is "100 square miles surrounded by reality" evoked an affirmative nod from most aspartame reactors who became involved. Several examples are listed.

- The quarterly FDA report on adverse reactions associated with aspartame consumption included the following data as of July 1, 1991.
  * It had received 5,903 complaints encompassing 91 symptoms.
  * Females constituted 77 percent of the complaints.
  * Headaches (1,463 cases) continued to top the list of symptoms.
  * There were 473 instances of seizures and convulsions by summating related catagories!
  * 308 consumers had a change of vision, including 4 cases of blindness.
  * Five deaths were listed under "other" problems!

- The FDA was asked to send a representative to the symposium on aspartame safety held at the University of North Texas on November 7-8, 1991. The organizers made this request so that interested persons could better understand the review process. The agency repeatedly refused to do so. Since the FDA maintains an office in Dallas, this educational assist should not have been an inconvenience.

Reaching the Boiling Point

Consumers and consumer advocates became exasperated as the number of severe reactions attributed to aspartame products continued mounting. The FDA already had logged over 5,000 complaints volunteered by consumers as of 1989, including more than 250 reports of convulsions! The updated data appear above. The reflexive response by the FDA concerning these "sadistics" remains one of nonchalance. This attitude recalls the remark by former Coach Henning of the Atlanta Falcons: "Statistics are like loose women. You can do anything with them."

Critics of the FDA within this realm have not been confined to concerned consumers, physicians and investigators. Their ranks swell as the Agency etches more emotional acid into the fine print of the Federal Register every time it permits the extended the use of aspartame. As far back as 1969, Dr. Herbert L. Ley, a former FDA Commissioner, stated

> "The thing that bugs me is that the people think the FDA is protecting them. It isn't. What the FDA is doing and what the public thinks it's doing are as different as night and day."

Vox Populi

Comments written in my questionnaire (see 22) attest to the intense anger of affected consumers concerning the FDA's entrenched attitude about the alleged safety of aspartame products. These correspondents resented its continued approval of such products, the inadequate or dubious "scientific" studies on humans, and the absence of proper labeling other than reference to phenylketonuria. (I would venture that 99 percent of the population is totally unaware of PKU and the likelihood they might be "carriers.")

Anger over the improper handling and labeling of aspartame products in the United States (see 26) also encompasses (1) failure to indicate the amount of aspartame in them, (2) failure to stress the need for avoiding prolonged storage and exposure to heat, and (3) absence of a printed expiration date. This general state of affairs caused nine states to form a "food police" task force charged with addressing the problem of misleading labels and ads (see 22).

Here are some plain-talk samplers from my correspondents.

• "Animals cannot speak and tell you how they feel."

• "Your FDA letter stated, 'This sensitivity to aspartame is presumably similar to the adverse response experienced by some individuals to various other foods and color additives, as well as to certain foods.' Well, somebody <u>presumed</u> wrong! I have never experienced the same problems with anything else in my life, nor have these people with which I'm familiar. It is very strange how you are trying to rationalize this problem, rather than believing the people and removing the problem."

• Dear Dr. Roberts:

"I have read your article in the St. Pete (Petersburg) Times on aspartame sweeteners.

"I too have complained, and was told 'there's no reason for the problems.' This was one year ago. They informed me that I couldn't get dizziness, nausea, the severe headaches, and the weight gain in just one week of using this product.

"I did a test myself. For one week I drank aspartame products, used the packets in coffee, iced tea and soft drinks. I used it in cooking as well as in cakes and cookies.

"After using this product, my attitude changed. I became irritable, cranky and fat.

"Then I contacted my general practice doctor. He said to discontinue its use (because) he got the same way.

"So, the second week I had everything in regular sugar or none at all. The symptoms have disappeared. I lost weight when I ate the same foods, prepared exactly the same way.

"I'm gonna fight this. If you need help, I'll help you! I wrote letters to the FDA awhile back, but no response. So, what do we do about it? Everything that's good has aspartame and I think it STINKS!!"

Encounter By a Consumer Group

The following is an unpublished excerpt (reproduced with permission) from a June 30, 1990 letter by Mary Nash Stoddard, President of the Aspartame Consumer Safety Network (see 36), to the <u>Wednesday Journal</u>.

"Far from being 'most respected regulatory agency in the world,' the FDA has in the past approved DES, the Dalkon Shield, faulty heart valves reputed to have killed heart patients, and was recently involved in the tawdry generic drug scandal reported by national media in 1989. This is the regulatory agency that nearly devastated the fruit growing industry of Chile after <u>reportedly</u> finding 3 mildly-contaminated grapes! (Protection of the citizenry or PR to rescue an agency's tarnished reputation?)

"W. J. readers might appreciate knowing that the head of the FDA's Adverse Reaction Monitoring System for Aspartame complaints met us in Washington (8/28/87) with a diet cola in her hand and a defiant attitude. On the conference table in front of her was a book edited by the highly-paid Researcher/ Consultants to NutraSweet. Through these insensitive actions while meeting with a group of aspartame consumers who had sustained injuries from the products (including blindness and grand mal seizures), representatives of the FDA showed their true colors. Absolute corporate neutrality on the part of a protective agency supported by <u>our</u> tax dollars? We think not!"

Discovery of the Heaviest Element

An article by Thomas G. Kyle of Los Alamos, "Heaviest Element Discovered," appeared in the Journal of Irreproducible Results (Volume 35, No. 1, p. 31). The attributes of this (fictitious) element, discovered by a (fictitious) Dr. M. Languor who termed it administratium (Ad), could be projected onto the not-so-fictitious FDA.

The author noted that Ad has several remarkable characteristics. The absence of electrons or protons resulted in an atomic number of zero. The element has one neutron, 75 associate neutrons, 125 deputy associate neutrons, and 11 assistant deputy associate neutrons — giving it an atomic mass of 312. Here are other impressive aspects of Ad.

- A small amount lengthened one reaction normally completed in less than a second to over four days.
- The half-life of Ad was about three years. But rather than decay at that time, Ad underwent an internal reorganization wherein all the associates, deputy associates and assistant deputy associates merely exchanged places.

"No Hard Feelings"

I was invited to address the 1990 annual meeting of the American College of Legal Medicine on the public health and legal implications of reactions to aspartame products.

The chairman for that particular session happened to be a ranking official in the FDA's drug division. Before this meeting began, I indicated that I would be making a few disparaging remarks about his agency. Adding that he shouldn't take them personally, I said, "Even if we disagree on this particular issue, we ought to remain friends." (The same reply has been used by my wife, an elected official, when she votes against some proposal submitted by a friend.)

My colleague's reply was reassuring. "Don't worry! You won't hurt my feelings one bit! Furthermore, I'm 25 miles away from that division."

The Senate Hearings

Three formal hearings concerning the alleged hazards of aspartame have been held by the United States Senate — two in 1985, and one in 1987. The transcribed proceedings will provide interested readers with extraordinary insights concerning the <u>modus operandi</u> and philosophy of the FDA.

> At the November 3, 1987 Senate hearing, FDA Commissioner Frank Young made the disturbing statement that he regarded money spent on aspartame research as "excessively disproportionate"! He then referred to the FDA's "poor facilities" for evaluating this chemical and products containing it.

Dr. Young's remarks about the "excessively disproportionate" amount of money spent on aspartame research were reinforced by Dr. F. Xavier Pi-Sunyer (1988), representing the American Diabetes Association's Board of Directors. He objected to this Senate hearing on the grounds it might make the public needlessly apprehensive!

Cartoon 42-1
© *1990 Mr. Sidney Harris. Reproduced with permission*

> "We have no indication from these professionals that there are significant problems with the use of aspartame...We do not believe, however, further studies are needed to duplicate current knowledge."

I promptly recalled the last remark on seeing another great cartoon by Sidney Harris in <u>Science</u> Magazine. (Cartoon 42-1)

The "Lackey-Handmaiden" Perception

Variations of the contention that the FDA serves as a lackey for Big Business repeatedly appeared in my correspondence from aspartame reactors.

> "Where is our FDA? The only winners are the impressarios who sit back and count their money...Sometimes when I think of the suffering it has caused me and countless others, I could tar and feather each and every one."

Consumers who felt they had suffered reactions to aspartame products realistically anticipated the futility of contacting this agency. One wrote, "I thought of writing to the company or the FDA, but figured it would do no good."

A concomitant theme pertained to the matter of <u>conflict of interest</u>. (One definition is "a conflict between private interest[s] and the responsibility of persons or agencies in an official position of trust.") To feature writers and cartoonists who had latched onto this issue, this was "government business as usual."

> • During the August 1, 1985 Senate hearing, Senator Howard Metzenbaum referred to the FDA as "...more of a handmaiden of the food and chemical industry than it is a defender of the health and safety of American consumers."
>
> • The laboratory chief of neurochemistry at the National Institutes of Mental Health denied serving as "a consultant to NutraSweet/Monsanto." He did admit, however, to being a scientific advisor in review-

ing pertinent research grants for International Life Sciences Institute-Nutrition Foundation, an organization supported by the NutraSweet Company (Nutrivoice Summer 1989).

An editorial by Dr. John B. Thomison in the Southern Medical Journal (Volume 83, 1990, pp.267-269) focused on yet another FDA pursuit: monitoring the generic drug industry.

> "What has been known less long is that there is in the testing, and possibly in the manufacture as well, a certain amount of cheating — how much is going on right now, and probably forever will, is anybody's guess, though the FDA is doing its best to find out. *(One hopes it really is doing its best, and that its best is good enough, since the FDA has itself developed a fair amount of tarnish on its escutcheon.* The tarnish has apparently been there a while, maybe even a very long while, and has been *discovered only just now because the public has up to now been blinded by the backlight of trust, as it turns out misplaced, in a government agency,* pointing up a lesson we should have learned a long time ago)...What it actually proclaims to the world is that we value human life when it suits us to value human life."*
>
> (Italics supplied)

The Search for a Regulatory Medical Diogenes

The March 15, 1990 edition of Nutrition Week contained an interesting announcement by the Assistant Secretary for Health in the Department of Health and Human Service (HHS). It concerned an opening for a new FDA Commissioner. The job description called for a physician or someone with an equivalent degree who could "protect the public

---

* *1991 Southern Medical Association. Reproduced with permission.*

health of the nation." (One of our best FDA commissioners was a Ph.D.-biochemist.) The annual salary depended on experience.

I admit to being intrigued. In effect, many of my studies and publications over four decades specifically involved this realm. I also was revulsed at the style of previous regulators whose wining and dining by industry had made the front pages.

The newsletter's editor inserted this unofficial proviso: "The candidate must be willing to make decisions, and willing to adjust them to the political objectives of the White House." For Yours Truly, such a tongue-in-cheek warning suggested prudence... and putting the idea of applying on the back burner. This coincided with chancing upon a provocative cartoon in which a physician was confessing to his nurse that he couldn't remember whether he had taken the Hippocratic oath or the bureaucratic oath.

Several weeks later, a professor of pharmacology at Johns Hopkins Medical School concurred with my decision after asking his advice. At home, however, the consensus was not unanimous. One of my sons urged: "Go for it, Dad! You're just the person the country needs in that job!"

> In retrospect, I'm glad I didn't apply because Dr. David Kessler was appointed in December 1990. I respected the reassuring no-nonsense approach of this physician-attorney at the outset, and wish him well.

Reservations About New Food Labels

Many American consumers anticipate clear and definitive information about the wholesomeness and safety of foods resulting from efforts by the FDA and USDA. In fact, these agencies have targeted May 8, 1993 as the date on which such a quantum leap in nutritional knowledge will occur due to the mandated changes in food labeling.

I wondered why none of my nutrition-oriented patients and friends expressed enthusiasm over the matter...but not for long. The bottom line was simple: "I don't trust those bureaucrats!"

> Irma Bombeck, noted humorous columnist, expressed a similar sentiment (The Palm Beach Post December 24, 1991, p. D-3). Even though she re-

garded nutrition as important, Irma just couldn't manage to dissociate this forthcoming achievement in labeling from other major bureaucratic brainstorms. They included "simplified" tax return forms, and the mandating of 29-cent stamps <u>before</u> being printed. In point of fact, Irma hoped that the few labels reading "NO!" could be a clue to locating foods still having a flavor.

AS-IS: A Déjà Vu

<u>Question</u>: "What do breast implants and aspartame reactors have in common at the FDA?"   <u>Answer</u>: "Comparable numbers of complaining consumers."

The FDA's medical device reporting system received about 6,400 reports of problems with breast implants over seven years by 1991. There were 3,397 involving silicone-gel implants, and 2,995 involving saline-filled implants (<u>The Palm Beach Post</u> January 5, 1992, p. A-18). By coincidence, the totaled figure approximates the number of persons who attributed significant medical problems to products containing aspartame (see above).

The frustration of both groups with FDA attitudes contained striking similarities. Such overlap was succinctly stated by Janet Van Winkle, founder of American Silicone Implant Survivors Inc., or AS-IS: "We don't exist. We don't have these problems." Hopefully, members of these organizations won't have to endure the 20-year somnolence of another Van Winkle, Rip, before an FDA boss wakes up.

Disinclination About Disinformation

Woe to <u>any</u> pharmaceutical company that might have fudged on a morsel of data, or uses even a <u>single</u> word to promote a drug that purists at the FDA consider misleading! The impact of such ire is awesome. It has been repeatedly evidenced by (1) apologetic "Dear Doctor" letters, and (2) full-page disclaimers in both the medical and general press costing hundreds of thousands of dollars.

A recent example of alleged "false and misleading advertising" involved the promotion of Naprosyn® at medical education symposia and in doctor-oriented cable television programs. The FDA raised a stink because of reference to this drug as being "arthroprotective" (i.e., reducing joint deterioration) in patients with osteoarthritis. Under a consent decree of permanent injunction, Syntex Laboratories, Inc. agreed on October 10, 1991 to inform physicians that "scientific studies" failed to confirm the arthroprotective action of Naprosyn. Furthermore, it would place similar notices, as paid ads, on Lifetime Medical Television and in 18 major medical journals.

All practising physicians also are "encouraged" by the FDA to report "serious reactions, observations of events not described in the package information insert (of drugs), and reactions to newly marketed products of particular importance." They are furnished copies of Form FDA-1639 in the FDA Medical Bulletin, along with detailed instructions.

In view of the Agency's concern about drug safety, some questions should be applied to aspartame:

- Has comparable accountability about overt reactions to aspartame products, or criticism about disinformation thereon, been leveled at manufacturers and producers by this Agency?
- Has it reconsidered the fact that aspartame is a drug, as initially conceived? (p. 176)
- Did the FDA go back to the drawing board in the face of 5,903(!) consumer reports of adverse reactions to such products received by July 1, 1991 — including 473 persons with seizures and convulsions, according to the total of pertinent entries?

If I've missed some of these details, Dear Reader, please relay them to me.

The same reservations apply to statements made by FDA officials. For example, The Palm Beach Post (December 10, 1991) carried a note of concern over "the amount of aspartame that could be consumed by 2- to

5-year-old children" in its Health Notes. The source: Dr. Linda Tollefson, a veterinarian and Chief of Clinical Nutrition Assessment at the FDA. Concomitantly, she made this paradoxic assertion, "We have never found a problem." But I found enough problems to devote an underline chapter to reactions among infants and children in my first book on aspartame reactions. They included convulsions, headaches, rashes, asthma, anorexia and gastrointestinal problems (see 38).

More on Corporate Disinformation

The matter of corporate disinformation is illustrated by a November 15, 1991 letter from the Senior Medical Consultant of The NutraSweet Company to Patricia Ryan, editor of Ideas Today. An article titled, "Diet Soda Issue: Do you Know What You're Drinking?," appeared in its September 1991 issue.

The corporate defender charged it "helped perpetuate the myth that unanswered questions remain about the overall safety of aspartame." He offered these "facts" (discussed in other chapters) as evidence. My responses follow.

- "The body cannot distinguish whether the components come from aspartame or common foods."

  It can...and does, especially in metabolizing its uncommon amino acid stereoisomers and other breakdown products (see 26).
- "The safety of aspartame has been well-documented by numerous controlled studies in both adults and children, as well as special subpopulations."

  In my opinion, the protocols for the vast majority of these studies were flawed —most notably through the ingestion of aspartame in forms that are vastly different from "real world" products (see Overview).
- "A number of well-controlled scientific studies by experts in the field of behavior indicate that aspartame is not associated with any changes in mood, cognition or behavior."

  My response: see above.
- "Many commonly consumed beverages and foods

also yield methanol upon digestion. A serving of tomato juice has approximately six times more methanol than the same amount of diet soda sweetened with aspartame. A medium-sized banana contains about the same amount of methanol (20 mg) as a 12 oz. diet soda. It would be impossible for a human to consume enough products containing aspartame to approach levels of methanol known to be associated with adverse health effects."

The fallacies of these assertions are discussed in Chapter 26, my previous text and several published articles. Such foods and beverages release only <u>minute</u> amounts of <u>free</u> methanol in conjunction with ethanol and other substances that have toxicological buffering effects.

The averaged daily intake of methanol is less than 10 mg. Whereas fresh orange and grapefruit contain 1-2 mg methanol per daily consumption, aspartame colas and orange drinks average 17,000 mg per 1,000 calories, and 112-182 mg for two liters (W. C. Monte: Aspartame: Methanol and the Public Health, <u>Journal of Applied Nutrition</u> vol. 36:43-54, 1984).

• "A retrospective analysis of the first 500 of the complaints was conducted by the CDC in 1984; the report stated that it was not possible to identify any 'specific constellation of symptoms clearly related to aspartame consumption.'"

Even a cursory analysis of this report would lead one to challenge such an interpretation of the data, and the purported insignificance of the "anecdotal" consumer complaints.

• "The FDA has conducted an exhaustive, ongoing review of more than 4,000 complaints in the 10 years since aspartame was first marketed (1981), which incidentally represents a small fraction of the 100 million individuals who regularly consume aspartame-containing products. The FDA arrived at the same conclusion as the CDC. Dr. Tollefson, of the FDA's Center for Food, Safety and Applied Nutrition, has stated that the adverse reactions allegedly related to aspartame, which have been received and reviewed by the FDA, '...do not establish reasonable evidence of possible public health harm. There is currently no consistent or unique pattern of symptoms reported with respect to aspartame that can be causally linked to its use.'"

My comment: see the Overview chapter and the foregoing recent statement by Dr. Tollefson.

• "The exhaustive aspartame testing that was conducted before its approval for the marketplace has been reviewed by the FDA and the U.S. General Accounting Office (GAO). Both organizations agreed the testing had been thorough and conducted properly. Hence, the article's suggestion that results of product testing submitted by G.D. Searle & Co. were falsified is quite untrue...GAO believes that these crucial studies met the FDA's requirements for the types of studies needed for a food additive approval."

That's not the thrust of the 1987 GAO report (GAO/HRD-87-46). It issued this disclaimer: "We do not evaluate the scientific issues raised concerning the study used for aspartame's approval or FDA's resolution of these issues, nor did we determine aspartame's safety. We do not have such scientific expertise."

Furthermore, 12 of the 69 scientists who responded to a GAO questionnaire had "major concerns about aspartame's safety." This report also cited the belief by the 1980 Public Board of Inquiry that "the aspartame studies did *not* conclusively show aspartame did *not* cause brain tumors" (italics by writer).

Thank You...But NO!

My concern over the possibility of aspartame-induced brain tumors (see 34) prompted this letter to Dr. David Kessler, Director of the FDA (see above), dated January 21, 1992.

> Dear Dr. Kessler,
>
> "I am writing you about the thorny issue of reactions to aspartame products. This unusual correspondence reflects my admiration for your efforts to date in fulfilling the mission of the FDA. Therefore, please regard it as a constructive letter from a colleague.
>
> "I have studied this matter rather extensively, as reflected by the number of articles and text I have written thereon. In addition, I communicated with several members of your agency concerning its data base on reported alleged aspartame reactions.
>
> "I am enclosing a just-published article titled, "Does Aspartame Cause Human Brain Cancer?" In essence, there is reason to project the demonstration of an increased incidence of brain tumors in rats during prior studies onto the recent increase of several major brain cancers in man.
>
> "I would be the first to acknowledge the limitations of this manuscript. The subject poses a momentous problem, however, that will require the input of a corporate-neutral investigator. I would like to collaborate with you in this capacity, along with statisticians on your staff and perhaps National Cancer Institute epidemiologists, rather than working on the project autonomously.
>
> "There is another pressing reason for you to take an active interest in aspartame reactions. The enormity of this problem has been downplayed by your staff --

even though it constitutes about 80 percent of complaints about foods and additives directed to the FDA
(FDA Consumer, October 1988, p.19). Accordingly,
the impression exists that the FDA is pro-business
rather than pro-consumer.

"Please let me know your thoughts on this possible
project."

I received a 6-page reply from Linda Tollefson, D.V.M., M.P.H.,
Chief of the Clinical Nutritive Assesment Section (see above), dated
March 20, 1992. The first and last paragraphs are reproduced.

Dear Dr. Roberts:
"Your letter dated January 21, 1992 to Dr. Kessler in
which you express concern regarding the safety of
aspartame has been referred to me for reply. Thank
you for your interest in this issue and for submitting
the article, 'Does Aspartame Cause Human Brain
Cancer?'

"FDA appreciates your concern regarding the safety
of aspartame and your desire to collaborate with
FDA scientists on the possible association between
aspartame consumption and brain tumors. However,
at this time we have no resources with which to con-
duct such research.."

This pro-aspartame apology contained more than a dozen statements requiring challenge. One conclusion became obvious: my further efforts to convince the FDA would be futile and a waste of energy.
Dr. Kessler and his FDA colleagues added an ironic twist in their
feature article about the safety of foods just before the printing of this
book. They emphasized in Science (vol. 256: 1747-1748, 1992) that the
FDA can remove any food from commerce "if there is even a 'reasonable possibility' that a substance added by human intervention might be
unsafe."

43

"FAKE FAT" RIPPLES

My interest in this field of nutrition had surprising ripple effects. They extended to "fake fats" (or "fat extenders," a less inflammatory designation), and other "imitation," "artificial" and "synthetic" edibles.

Initially, I was reluctant to get involved, largely due to concern over the accusation of pursuing some corporate vendetta. I hereby go on record that this has never been a factor.

My professional conscience, however, had been pricked. Having documented the ordeal of so many persons with reactions to products containing aspartame, I couldn't ignore the possibility of subsequent tidal waves involving side effects to the many unnatural substances (xenobiotics) Madison Avenue already was christening with impressive "lite" names. The fact that they had not been subjected to extensive trials in humans before approval became another variation on a familiar theme.

> Again, Ogden Nash had beat me to the punch. He astutely observed that we often seem to be making great progress...but in the wrong direction.

Act One

I submitted two related public comments to the FDA during 1989. They recommended delay in the approval of Simplesse® pending convincing answers to several basic scientific and clinical questions. My "Open Letter to the FDA," published in the Journal of Applied Nutrition (Volume 41, Number 1, 1989, pp. 42-43), also pleaded for restraint.

Jeff MacNelly

As these events unfolded, I had the eery feeling that Jeff MacNelly, one of my favorite cartoonists, either was looking over my shoulder or had been working toward an advanced degree in nutrition.

Shortly after submitting my first public comment, Jeff's SHOE strip focused on a shopper who was frustrated after encountering so many "fake" products in a market. They included "the greatest invention of them all: 'FAKE FAT!!'" He then informed his sidekick what makes our country great: "Real choices."

SHOE   by **Jeff MacNelly**

Cartoon 43-1

The heroine of <u>Cathy</u>, a comic strip by Cathy Guisewite, also bemoaned living on "fake sweetner...phony chocolate...butter substitute" (<u>The Miami Herald</u> July 15, 1990). She begged for an artificial fat that hadn't even arrived at markets.

I cite this strip for its more profound message — namely, an undermining effect on the collective female self-image relative to "demeaning our expectations for our lives" by pursuing and consuming imitations of the real thing.

Act Two

The following commentary by Yours Truly appeared in the <u>American Medical News</u> (January 12, 1990, p.32), a newspaper for physicians.

Not enough data on 'fake foods'

"Your feature on Olestra and Simplesse (AMN, Sept. 8, 1989) requires additional perspective for physicians. These and other 'fake foods' and 'designed additives' are likely to radically alter the American diet. Unfortunately, licensure for products made from unnatural substances classed 'generally recognized as safe' (GRAS) is being sought without sufficient biophysiologic data, and in the absence of extensive pre-marketing trials on humans.

"I have submitted public comment on these issues to the FDA, seeking restraint before approving Simplesse. It will appear under 'GRASP 8GO345 — Microparticulated Egg and Milk Product (Simplesse).' To clarify my status, I am a practicing board-certified internist. I have received no financial support from any vested corporate interests. My concern evolved from the frequency with which I have encountered reactions to aspartame, as reported in multiple publications.

"Among the unanswered questions about Simplesse: Where are the data by independent peptide chemists on the stereoisomer concentrations of the amino acids, especially following processing and storage? Where is the information after post-absorption concerning its metabolism and entry into the central and peripheral nervous systems and the lens, and the effects on neurotransmitters? Can this product induce malabsorption and nutritional deficiencies other than the excessive caloric reduction sought by weight-conscious persons? Can it undermine the nutrition of young children?"

Act Three

The "other shoe" (not Jeff MacNelly's cartoon character) dropped on February 22, 1990 when the FDA approved the "all-natural" Simplesse®. Would-be consumers salivated over 'the prospect of guilt-free Simple Pleasures®. The manufacturer also salivated as Monsanto's stock soared $4.37 that day.

Cartoonists have taken note of the "culinary guilt factor" underlying these responses. In the March 30, 1990 issue of <u>Science</u>, Sidney Harris sought to alleviate the sense of guilt inflicted by vegetarians on meat eaters. His cartoon depicted two lions finishing the remains of a four-legged victim as one uttered, "...a good meal...and no guilt."

Not being aware of any new data concerning the metabolism and tolerance of this product in humans, I sent the following letter to Senator Bob Graham of Florida.

"I am dismayed and alarmed — as a practicing physician, medical researcher and concerned citizen — over the approval of Simplesse by the

FDA. This action will result in millions of Americans becoming experimental guinea pigs. To my knowledge, no widespread studies had been done on <u>humans</u> to confirm the 'generally recognized as safe' (GRAS) status of this 'fake fat.'

"The FDA decision raises the grim spectre of a repetition of the ongoing intense controversy concerning the safety of aspartame. The details appear in my recent book, <u>Aspartame (NutraSweet*):Is It Safe?</u> (The Charles Press). The reasons are much the same — namely, the unknown ability of the human body to handle high concentrations of amino acids that have been profoundly altered through manipulations by food technologists. One must not be misled by the glib assertion, 'An amino acid is an amino acid.'

"Hopefully, my anxieties on this matter will prove unwarranted. But I have yet to be provided with any of the answers I requested concerning basic scientific questions in my formal public comment to the FDA challenging the premature licensing of this product."

But, at last, I was in synch with Jeff MacNelly — specifically, his strip on February 26...<u>the very same day</u> this letter was mailed. My letter now could supply SHOE with something in the news he could "view with alarm."

Cartoon 43-2

Act Four

As night follows day, a sugar-free version of Simple Pleasures® was destined to make its appearance on the synthetic food front. It did so in the form of Simple Pleasures Light®. The sweetener? You guessed it: aspartame. The flavors? Vanilla, chocolate, vanilla fudge swirl, and chocolate caramel sundae.

Concomitantly, McDonald's announced the availability of its McLean Deluxe Hamburger. It contained 90 percent meat, 9 percent water, and one percent seasonings — including carrageenan (a powdery white vegetable gum from seaweed for retaining moisture).

One food columnist now envisioned the dieter's utopia: a carrageenan-containing burger with diet soda, followed by fat-free cake with a scoop of fat-free, sugar-free frozen dessert a la mode (The Washington Post March 20, 1991, p. E-3).

Matzo Balls: Another Form of Microparticulated Egg

My wife holds an unofficial title locally as The Matzo Ball Queen. It's a long story that even has political overtones. Suffice it to say, anyone who has sampled this "specialité chez Roberts" will praise Carol's culinary skill. (She also was the Gastronomique of our chapter of Confrerie de la Chaine des Rotisseurs, a gourmet society.)

So what does this have to do with Simplesse? By way of orientation, the product is formally described as a "microparticulated egg and milk product."

I will now reveal an important "secret" to Carol's recipe for success-ful matzo balls: rapidly beating the eggs by hand for five minutes. This introduces just enough air for the balls to expand and maintain the proper "lightness."

As this book neared completion, Carol recruited Yours Truly to lend such a hand...literally. I obliged. Dutifully performing the chore, a thought struck, "Why! I'm producing another microparticulated egg product!"

Correspondence

The "fake fat" theme increasingly kept insinuating itself in my corre-spondence. For example, the enlightened executive director of a na-tional nutrition organization described the devastating consequences of very-low-fat diets in some members. They included gallbladder attacks, and failure to thrive by the child of a woman who avoided fat during pregnancy. She summarized her attitude in this equation that could have made Einstein take notice

FAKE FAT + FAKE SUGAR + FAKE CREAM + FAKE BUTTER + FAKE PROTEIN = FAKE PEOPLE

44

WHO KILLED THE SUGAR PLUM FAIRY?

A Very GRIM TALE Indeed As Told to Mary Nash Stoddard
by the Ghost of Mother Goose

Preface By The Author

The following "fairy tale" is the brain storm of Mary Nash Stoddard, President of the Aspartame Consumer Safety Network (see 36). It is reproduced, slightly modified, with her permission.

Conventional fairy tales and nursery rhymes have long influenced the public's sweet tooth. Some undoubtedly enhance the "fear of fat," such as those dealing with "sugar and spice" (see 24).

I also mentioned other relevant fables and myths in previous chapters — e.g., the emperor without clothes (see 5).

A few authors utilize "fractured fairy tales" to provide perspective. Jon Scieszka's The True Story of The 3 Little Pigs (Viking) provides an example. This elementary school teacher tried to convey the message that there are at least two sides to every story.

- Was the wolf at fault for not wishing to let a fresh ham dinner go to waste?
- Who was to blame because a pig failed to use proper construction materials when building a house?
- And even more germane, what about the old woman who lived quietly in her forest gingerbread house until beset by two annoying children named Hansel and Gretel? (See Cartoon 4-2)

Fairy tale themes even surface in the attitudes of sophisticated young adults and corporate executives.

- The Wall Street Journal carried a full-page Toyota ad that headlined the sentiment of Karla Thomas, Miss Bethune-Cookman College for 1989/1990. It read:

"I'd like to thank my Mother, my Father, and especially my Fairy Godmother" (October 26, 1990, p. B-8).

- Reference was made to <u>Alice in Wonderland</u> in Chapter 26. It continues to be an attention-getter — e.g., an ad by the Mobil Oil Corporation titled, "Alice in Blunderland."

Cartoonists have used this approach in presenting the other side of the aspartame story.

Bill Dwyer depicted an encounter with "Mindy Moonbeam" in Cartoon 3-2. She was a chap's favorite "fairy of wishful thought," who came to help him in "rationalizing drinking the diet soda."

Some courageous souls have felt the need to modify or update classic fairy tales because of their sad endings or sexist implications.

- Mike Royko, the syndicated columnist, reported on the valid criticisms of old fairy tales, such as <u>Cinderella</u> and <u>Snow White</u>, by Chicago school teacher Georgiann Carlson (<u>Sun-Sentinel</u> October 10, 1990, p. A-11). She regarded them as sexist exercises that undermined the independence and self-worth of young women through emphasis upon the some-day-my-prince-will-come theme. So, Ms. Carlson rewrote these tales, and gave them different endings. The results: Snow White rescues the prince, and Cinderella gives her prince the brush-off to marry a stable boy.
- Retired obstetrician Herman Kantor wrote <u>Mother Goose & More</u> (Additions Press) because of a decades-old compulsion to provide young listeners and readers with happy endings to morbid or dangling rhymes. Accordingly, he patched up Humpty Dumpty, found biscuits for Old Mother Hubbard's

hungry dog, saw that Jack Sprat gained some weight, and provided domestic intervention for the Old Woman Who Lived in the Shoe. "Dr. Hickey" (so nicknamed because of his H.I.K. initials) commented on the latter: "It's a sad poem, whipping them all and sending them to bed. That's not how you do it with children. *We had to sweeten it*" (The Miami Herald November 23, 1990, p. C-4). (Italics supplied)

Mary had no way of knowing the extent to which invent-a-virus scenarios were being concocted in scientific circles.

Leo Kinlen, an epidemiologist at the University of Edinburgh, proposed the "new town hypothesis" in Science (Volume 248, April 6, 1990, pp.24-25). It averred that clusters of leukemia around nuclear plants might be viral in origin, especially among workers who were new to the area and lacked immunity.

Getting back to the world of cartoons, Peanuts often managed to get in the last word. For example, Lucy asked Charley Brown why he didn't look happy. Charley replied with the very same are-there-side-effects question I've been asking about persons who consume considerable aspartame products. Too bad it hadn't been answered in the following case involving Sweetener Plum Fairy.

*   *   *   *   *

Once upon a time, in a land far, far away (not this one, of course), a little old chemist deep in the darkest part of the forest was toiling over a hot beaker...trying to find a good medicine to cure bad tummyaches. All of a sudden, the noxious mixture boiled over the cauldron — I mean beaker — and spilled onto his finger. Later, he licked his finger and it tasted sweet. Very sweet. Yummmmmmmmmm. [See Chapter 28 on sweetness and serendipity.]

Because it was very sweet, he knew everyone would love it. For, you see, he lived in the land of the Sugar Plum Fairy. She was so beautiful and sweet. Everyone worshiped her because she was so sweet. They taught their children to worship her sweetness. In fact, they all loved sweetly worshiping the Sugar Plum Fairy and the great TV-god. No matter that they were all becoming fatter and fatter from all those sugary confections, and could not climb two consecutive flights of stairs — even if their lives depended on it — because they watched too much TV.

This was a land filled with people who were "ripe" for a sweet new product on which to gorge themselves and their children.

The company for which the little old chemist worked was named W. E. Swirl Inc., listed on a "big board" as WESINC. The family who owned the company rejoiced and rejoiced when they heard the good/bad news. This meant that they would not go broke after all. Actually, this bit of good/bad news conjured up visions in their heads of a product that would go on and on, and earn billions and billions of Sugar Plum Fairyland dollars.

But alas, dear children, a problem arose. All the people in Sugar Plum Fairyland would have to be told that the Sugar Plum Fairy, whom they all loved and worshiped, was really evil and bad. How could they get the people to believe them?

This is what they did. They invented something new for the people to worship, and to teach their children to worship. They called her the Sweetener Plum Fairy. But first, they had to fool the people and make them think she was good for them — like making them skinny and fit. And, would not hurt them...no, not even one little bit.

Before the Sweetener Plum Fairy could be presented to all the people, however, the company had to get approval of a special governmental agency in the capital city of Fairyland. To get permission, it must prove to this Auspicious Government Administration (A.G.A.) that the good/bad new Fairy would harm no one in the sweet kingdom.

That's when things started to go wrong. Oh, dear me! Swirl's Sweetener Plum Fairy failed miserably when it was tested. She made lots and lots of little lab creatures sick — indeed, very, very sick. She even made some of them very, very <u>dead</u>.

If revealed, of course, the results of these tests would be ever so embarrassing. So, they weren't revealed. At least, not by WESINC nor by the A.G.A.

Now, where was I? Oh, yes. Swirl said that the bad test results must be covered up. So, the company's strategists really burned the midnight oil. (Yawn)

Finally, the strategists drew a plan. A new good/bad man would be hired to head the company, and help it cut through all that big and inconvenient red tape. So it chose a person with the highest political connections imaginable. How high is high? Just try to guess, boys and girls! Would this action set a president, er— I meant precedent, to deal with governmental agencies? Maybe.

Then, one fateful day in the year 1981, the Sugar Plum Fairy started to get sick...very sick. Two years later (pardon my tears), she died. The people mourned. Some whispered that the people of Swirl and the A.G.A. had killed her. But what does it matter now? The deed was done.

At last, WESINC could reap its vast — and I do mean vast — rewards. Things looked ever so rosy. Tra-la-la-la-la. Even the great TV-god was taught to sing the praises of the new, improved Sweetener Plum Fairy.

Now, I must tell you about a little boy with big, sad eyes. He was shushed by everyone in the room when he said, "The Sweetener Plum Fairy's swirly trade mark looks just like the picture of a big, bad storm." Everyone glared at him. Of course, he <u>never</u> said <u>that</u> again. Never!

Then, strange things began to happen in Sweetener Plum Fairyland. Some folks started to get sick from eating foods and drinks that contained the new sweetener. They began complaining to their doctors, and even to the A.G.A. Reports of alarming symptoms began to escalate.

But where, oh where, had <u>all</u> those terrible complaints come from? The brain trust of the A.G.A. thought...and thought...and thought. Then, it occurred to them: those poor little creatures who had been tested in the lab!

Ooops! "Not to worry," soothed the company. After all it had gone through to win approval of the A.G.A. and the people in Fairyland, there was no way it would let this best-selling new, improved Sweetener Plum Fairy collapse. No way! Furthermore, it was helping the economy. Most of the people and their children who consumed mega-gallons of the stuff were now wearing designer jogging shoes. And there were plans in the works for WESINC's sweetened dog food, motor oil, and even shoe polish!

Now, what do you think, boys and girls? Was Swirl's product recalled when the people started getting sick? Of course not! Was approval by the Auspicious Government Administration rescinded? Don't be silly! Instead, the company and the A.G.A. together decided on a just-dandy idea: extend approval of the Sweetener Plum Fairy stuff to more products.

But how about all those poor folks who were going to feel like all those poor little creatures in the lab whose tests were never reported by WESINC? Time to burn the midnight oil again. (Yawn)

Then, the people in charge of WESINC had a brilliant idea. "We'll set up an organization that is totally separate from our company (so that no suspicions will be raised) to fund researchers and doctors who will say..." Now, let me think, because this isn't easy to put in words. Okay, here's how it went:

> "We'll manufacture a fake epidemic, and pay people lots and lots of money if they'll say that a bad old virus has been going around Fairyland for a long, long time... certainly before the debut of our new, improved Sweetener Plum Fairy. And this virus causes the very same symptoms that those folks who have been eating and drinking things with WESINC's sweetener say they have.
>
> "Since more people in Fairyland will be using the swirly stuff in everything, and more people will be feeling bad, maybe our diversionary tactics would convince them and their doctors that they have this bad old virus, too. They could then form a Chronic Bad Old Virus (CBOV) Syndrome Association with its own support groups to let people comfort each other. We'll also pay a "yuppie scribe" to write a scary fairy tale about the dreaded CBOV syndrome, and get it published in Tumbling Rock (the favorite magazine of Fairyland's yuppies).

> "Finally, we'll even convince the great TV-god to feature stories about the CBOV on its news and talk shows, and programs like that."

You know what, dearies? (Here's where it also gets a little tricky, but try to bear with me.) The Association's <u>paid</u> staff members told some of these poor sick folks who called its hotline number that most patients with the CBOV syndrome began to feel better by changing their diet. Yes, that's it. "Do not eat sugar (gag), and go a bit easy with the swirly-sweetener."

Now, goodness me, weren't those fellows at WESINC ever so clever to think of all that? They concocted a creative way to remove any blame from their company, and transfer it elsewhere. I'll bet that scheme took lots and lots of (Yawn) midnight oil, kids.

Meanwhile, WESINC lawyers were jumping up and down from joy. They clapped and clapped their gnarly hands in glee. After all, they roared, "Who would dare <u>sue</u> (hee hee)...<u>A VIRUS</u>??"

All the while, the new sweetener had twirled and swirled out of Sweetener Plum Fairyland until it encircled the whole wide world. And what about the little boy with those big, sad eyes who said the swirly symbol looked like a big bad storm? He just watched quietly, but <u>never</u> uttered <u>that</u> again.

There are those who say that the Sugar Plum Fairy may not have been perfect, but neither was she all that bad. If only the folks in Fairyland hadn't worshipped her quite so much! Maybe she's only asleep and waiting to be awakened by a kiss from the Handsome Prince of O.U.C.H. (Organization of Undetected Consumer Hazards), whose arrival had been predicted in the <u>Journal of Irreproducible Results</u>. Just like Sleeping Beauty.

Now, wasn't that a fantastic story, children? Of course, I know it's soooooooo silly! Why, something like that could <u>never</u> happen here in America. Just trust old Mother Ruse...I mean Goose.

*   *   *   *   *

"But wait, Mother Goose! This is the Old Woman Who Lived In A Shoe speaking. And don't go to sleep just yet, kiddies, 'cause I have a <u>very</u> similar story to tell you...and this one happens to be true."

45

"TO HELL AND BACK ON ASPARTAME"

Preface By The Author

I received this case history, including the very same title, in June 1990. It was not written by the Old Woman Who Lived In A Shoe (see 44), but by Kelly Allen of Gillette, Wyoming. Reference to her driving off a cliff was made in Chapter 38 — specifically, by her 9-year-old son during a creative writing workshop. This "story" is reproduced, with only minor editing, by permission of Kelly.

There's an important reason for including Kelly's experiences. "Specialty" or "gourmet" coffees sweetened with aspartame increasingly are being consumed, especially by young women. The same applies to the pre-sweetened iced coffee drinks made with vanilla extract and aspartame that are now available at the take-out counters and drive-in windows of nationwide chains. These "coolers" actually may be preferred because they avoid the bother of blowing on hot coffee or tea. And then there are the popular pre-sweetened teas, as noted below.

Other aspartame reactors also have used the term "Hell" to describe their misery. For example, a 39-year-old computer programmer suffered severe mental and physical symptoms after increasing his intake of an iced tea drink sweetened with aspartame. He precisely recorded them in his diary. Most disappeared within one month after stopping this product. He wrote, "My workout journal became an increasingly detailed diary of the Hell I was going through as I tried to come to grips with what was wrong with me."

Now, just try to project Kelly's "anecdotes" and "idiosyncratic reactions" onto the thousands or tens of thousands of her counterparts using "my coffee" or "my tea."

*   *   *   *   *

Introducing new products into society is important. However, side effects should be stressed, and possible interactions with other chemicals should be labeled big enough to see.

I am a 30-year-old female. I grew up in a happy home, 2 sisters, 3 brothers; all still alive and healthy. My parents were wonderful. Everything we ever attempted, they supported. If we needed help, they were there. It's still that way today.

I now reside in Gillette, Wyoming, with my husband and three sons. There are very few problems in our life, my health being the major concern for family and friends.

Being in and out of several medical institutions, I have been diagnosed as having different illnesses. Basically, I have been made to feel like "a nut." We have spent thousands of dollars trying to find out what the problem was, but with no one ever really coming to a conclusion.

They (the doctors) had almost convinced me that maybe there really wasn't anything wrong. Maybe I <u>was</u> a nut. Last November this changed. The problems were real and the symptoms were there, but we (the family) just couldn't tie everything together. There have been major problems along with some minor everyday occurrences.

The first major change I noticed was my head getting very heavy after I ate, sometimes to the point I would black out for a few seconds at a time. I wasn't eating much at this time, anyway. Everything I ate nauseated me. Several drugs were prescribed to help cure this.

I also was tested for hypoglycemia. They concluded this was my problem, and I was instructed what to do. I followed the guidelines fairly close.

One day, while making a delivery to someone at Federal Express, I took a peppermint candy from a dish they had sitting on the counter. Proceeding on my way, I picked up the kids from school and continued along. From there I'm blank. I felt like I was going to pass out, so I tried to pull off the main street. Upon regaining consciousness, we went straight home. Not one word was said.

When my husband came home that night, Andy, 7, ran to the front door and yelled, "DAD! Mom tried to kill us today! She was driving down the wrong side of the road. All the cars were coming at us. Then she drove in the ditch and across the grass!" My husband looked at me. I could not explain. When I came to, we were three blocks away from where we'd started. It's a miracle I didn't hit someone or something.

This was not the first driving incident that occurred. There were two others. In 1987, I drove off a 30-foot cliff after eating breakfast. In 1989, after being diagnosed as lactose intolerant, I tried something the nutritionist gave me. I ended up blacking out and driving up onto the railroad tracks. I had to be winched off.

It was time to prioritize. I gave my keys to my husband and didn't drive again. There comes a point when you have to worry about the other people on the road.

John Cougar Mellencamp's song, <u>Rumbleseat</u>, kept me going from there. I would play it over and over, convincing myself I could figure this out. My life was like playing Russian Roulette. Everytime I put something in my mouth, I would wait for the reaction. I was experiencing several different symptoms: nausea, dizziness, heavy head, blurred vision, terrible gas, bad breath (to the point I went to the dentist), loss of bladder control, menstrual disorders, confusion, and more.

These problems kept mounting. I made an appointment at the Mayo Clinic in Rochester, MN, for December.

It is hard to function normally with all these recurring symptoms. I have a nailcare business which I run out of my home. I had to ask my customers to start coming to me, as I was not driving.

So now Thanksgiving has passed, and I'm still headed for Mayo. While servicing a customer at home, we got to talking about all my problems and health in general. She said, "The funniest thing happened to my sister. She was driving down Burma when this lady drove right up on to the railroad tracks. They had to winch her off. My sister thought the woman may have been drunk, but there was no alcohol smell. She seemed really confused. My sister then thought she may be on some sort of drug because she was slurring her words. She said the woman just thanked them and left." I informed her that the person on the tracks was <u>me</u>, and to inform her sister that I had quit driving.

What should happen next????? I had been feeling pretty good for about a week and hadn't had any blackouts. I decided to run a few errands. It was hard to sit at home and wait for things to happen. I made myself a cup of "my coffee" (a famous-brand product with added aspartame), and was off. Before I knew it, I was feeling horrible. I had to rest at the eye doctor's office where I'd gone to make a delivery. When making the next stop, she asked how I felt. I was not doing well. I got in the car and prayed I would make it to my husband's office. I did. We went straight to the emergency room. My body was numb. Vision blurred. The left side almost seemed paralyzed. They began testing me. Again, they could find nothing wrong. I was told it was probably a migraine complication, and was sent home. They instructed me to keep my appointment at Mayo.

So now I'm back to getting rides, and my customers coming to me. My diet consists basically of Chinese food, McDonald's fish sandwiches, tuna for breakfast, and "my coffee." "My coffee" always made me feel better and seemed to give me energy.

Gene, my husband, thought it was the caffeine causing the problems. So I went to drinking diet pop. The gas problems got worse, as did my breath. Off to the dentist checking for cavities; none. He felt the bad breath was caused by some sort of stomach acid. Since I hadn't been eating any acids, no pizza for 2 years, no fruits, tomatoes, etc., we were at another dead end. I began eating Tums, about 15 a day, to help the gas problem.

Still no more foods into my diet, but I could drink "my coffee." It was one of those specialty brands. Whenever I would drink regular coffee, I got "the shakes."

Try functioning normally when everything you put into your mouth causes some sort of problem. If we were out and I got "the shakes," my husband would twitch an arm and leg and say, "M-Y W-I-F-E H-A-S B-R-A-I-N D-A-M-A-G-E." This would lighten the situation but it really hurt my feelings. Basically, everyone would shrug it off and go on with what they were doing.

So now it's December 1989. I'm starting to lose several pounds again. (In May of 1987, I lost 25 lbs. in a week for no apparent reason.) Back to the doctor again. They don't know. "Keep your appointment at Mayo," they said.

It's now time to go to Minnesota. We had to turn in our leased Bronco in Rochester. We'd already bought a Suburban here, with the idea that I'll be safe if I wreck again. I tried to drive to Rochester. I made it as far as Austin, but then had to stop. I was beginning to feel the heaviness in my head. We left the Bronco at a gas station. Gene and my brother-in-law went back for it after we got to their house.

The appointment at the Mayo Clinic. Several tests, several different doctors, and hundreds of dollars again...only to be told that all of my organs were healthy and I should live a long life. I did, however, have an abnormal EEG. (Their final diagnosis was "partial complex seizure disorder and grand mal seizure disorder secondary to a history of viral encephalitis.") This puzzled us since I already had several, and most were normal. In fact, after going to the Billings Clinic for the same sort of problem in '85, the doctor told me he didn't feel "the shakes" and other symptoms I was experiencing was a neurological problem.

So now what? The doctor at Mayo tells me to be positive. "Write down what you can eat, and write down what you can't, and eat what you can." He also put me on phenobarbital; I requested we try this since I'd tried several other medications and none of them worked. This was in the drug they gave me after my hysterectomy, so I felt it may work in a large dosage. Before leaving the Mayo Clinic, I did go eat for the nutritionist. By the time we (Gene and I) got back to the office, I was totally out of it. It was like being intoxicated. He acknowledged there was something wrong, but didn't know what it was. I had blood drawn. They asked what I'd been drinking. Nothing showed up. I left Mayo with little hope.

The next day, my mother was taking me. More tests, no answers. After our last consultation, I lost all hope. She could see the disappointment in me. We stopped for hot water. I made "my coffee" and we drove back home. My mom was trying to be positive since I'm supposedly healthy. I broke in with "but I can't eat or drive." When we got home we played cards for hours. All of my brothers and sisters tried to cheer me up. They waited on me, making "my coffee" as I needed it. We quit about 3:30 A.M. and I went to bed.

When I awoke the next morning, I went downstairs, made "my coffee," and sat to visit with my brothers. Mom was in the kitchen making a breakfast dish she knew I could eat. I ate it, and as we put it, became plastered. I could not talk right. Steven, my brother, looked at me and said, "Kel, you'd be a cheap date. We could order you a donut and you'd be set for the night." I lost it.

I was very upset and went to the sunroom to get away from everyone. I sat on the floor between a plant and the radiator. I was crying by this time. "What do I do now? Everything is rushing through my head. I _am_ a nut. I can't drive. I have brain damage. I can't eat or I'll get drunk."

My mom entered the room. She didn't know what to do. She'd never seen me this bad. We talked and she kept trying to reassure me. She focused on the positive things, the new house. I told her I would be dead within six months, and would die in my new house. She didn't know what to do, and left the room.

In comes Gene, trying to help. I told him I was tired of feeling like a weirdo. I couldn't deal with it anymore. I tried to get him to shoot me. I begged him, this being a recurrence of what happened after I wrecked the car in '87. Life was hell, and I wanted out! All that came to mind was, "What about the kids? Someone help me."

Everyone avoided me the rest of the day. Whenever I went into a room, everyone left. It was Steven's birthday, and we were supposed to go out. Steven decided he wanted "carry out" Chinese at home. Everyone ate. I was upstairs. I went down to eat after they'd all finished. Within an hour I was feeling better. Gene and I took the kids to see some friends. We stopped at the store so that I could get some more "coffee." They didn't carry my kind. I thought it was strange that they didn't have that one kind in Gillette, and now it wasn't in Minnesota either. I went home and asked my mom. She informed me that she had it ordered at another store. So I got it there.

Time to head back to Wyoming. Another Wyominite was planning to follow us. When we stopped for lunch, she asked if I'd ride with her for awhile. I did. At lunch, I had eaten things we knew I could eat. We began our trek through South Dakota. I began to slur my words and show signs of intoxication. She was getting scared. Within 30 minutes I was in a depressed state. We began tracking everything I had eaten, and concluded it had to be the "coffee" and the diet pop. I immediately stopped drinking them. I started feeling better. However, I wasn't sure if it was the medication or the beverages I had omitted. I drank some more "coffee," and started to get all the symptoms again.

Our family has been struggling through this now for 5 months. The kids are starting to believe their mom is okay. I am able to eat anything I want. I can drive. I <u>haven't blacked out since stopping "my coffee sweetened with aspartame."</u> It's like a new life has started for us.

Now, what about the other people who are living in this hell that we've been in? People need to be warned of the side effects these ingredients may have, as well as how they are going to interact with other things in their diets.

46

"CASE 552"

My final offering consists of another pertinent correspondence. It was written by Mary Lou Williams of Plainview, New York.

Mary Lou wrote a feature in the April 27th, 1990 edition of the Syosset Tribune titled, "Artificial Sweeteners: The Bitter Truth." It quoted from my article, "Reactions Attributed to Aspartame-Containing Products: 551 Cases," published in the Fall 1988 issue of the Journal of Applied Nutrition.

"Tracey's Story — Case 552," slightly edited here, followed this excerpt. In gratitude to Mary Lou, I designated Tracey as Case #552 in my series.

The editor of the Syosset Tribune subsequently received a lengthy "rebuttal" from Dr. Robert H. Moser of The NutraSweet Company (see 17). Her response reminded me of the little boy in Chapter 45 who was admonished for interpreting the "swirl" as a gathering storm.

*   *   *   *   *

My interest in aspartame was sparked by the experience of my niece, Tracey, about four years ago when she was twenty years old. She had been drinking diet soda containing aspartame for about a year. During that time, she was subject to bloating, which became more severe as time went on. She also developed a craving for sugar, which she had never had before. This craving led to frequent sugar binges.

As a result of the bloating and the bingeing, Tracey gained weight and decided to go on a diet. Her diet consisted of nothing but diet soda — two large bottles a day — for a week. At the end of that week, she had the first of a series of bizarre episodes. She fell to the floor and, although conscious, could not pick herself up, balance herself to walk, or even lift up her head. Her brother carried her to the hospital. The eyes were rolled back in her head. She had a very low pulse, very low body temperature, and very low blood pressure. She lapsed in and out of consciousness for the next eight hours.

The doctors could find no cause for these severe symptoms, and discharged her without a diagnosis. She regained full consciousness and the ability to walk unaided. She had continuing blackouts that were of short duration (one or two minutes), but frequent — about once a day.

Tracey had another severe attack two weeks after the first one. Her limbs shook uncontrollably; her eyes again rolled back in her head; she became extremely weak; and she hyperventilated. She was again rushed to the hospital. Again, they could find nothing wrong even though she lost consciousness for ten minutes in the emergency room.

My sister-in-law at this point suspected the diet soda because of the addiction my niece seemed to have for it. Tracey did not believe that was the cause. After all, it was presumably made from natural ingredients. But she was desperate enough to try anything...and stopped drinking the soda.

Tracey began to feel better and stronger. The bloating subsided. She lost weight just from not drinking diet soda. Her craving for sweets abated. The recent loss of consciousness and the severe headaches that preceded it no longer occurred. In fact, she has been fine ever since. The only exception occurs when, inadvertently, she eats anything with aspartame in it. Even a piece of gum will precipitate a severe headache.

PUBLICATIONS OF H. J. ROBERTS, M.D. ON REACTIONS TO ASPARTAME PRODUCTS

Is aspartame (NutraSweet®) safe? On Call (official publication of the Palm Beach County Medical Society), 1987; January: 16-20.

Aspartame (NutraSweet®)-associated confusion and memory loss: A possible human model for early Alzheimer's disease. Abstract 306. Annual meeting of the American Association for the Advancement of Science, Boston, February 13, 1988.

Aspartame (NutraSweet®)-associated epilepsy. Clinical Research 1988; 36:349A.

Complications associated with aspartame (NutraSweet®) in diabetics. Clinical Research 1988; 3:489A.

The Aspartame Problem. Statement for Committee on Labor and Human Resources, U.S. Senate, Hearing on "NutraSweet"—Health and Safety Concerns, November 3, 1987. 83-178, U.S. Government Printing Office, Washington, 1988, 466-467.

Neurological, psychiatric, and behavioral reactions to aspartame in 505 aspartame reactors. Chapter 45 in Dietary Phenylalanine and Brain Function, edited by R. Wurtman and E. RitterWalker, Birkhäuser Boston, Inc., 1988, 373-376.

Reactions attributed to aspartame-containing products: 551 cases. Journal of Applied Nutrition 1988; 40:85-94.

New perspectives concerning Alzheimer's disease. On Call (official publication of the Palm Beach County Medical Society). 1989; August: 14-16.

Reactions to aspartame (NutraSweet®). The Expert Witness Journal 1989; August: 1-3.

Sweetener feedback. (Letter) General Aviation News 1989; August 28: 4.

Report links aspartame to pilot and driver error. The Palm Beach Post October 14, 1989, E-2.

Not enough data on "fake foods." American Medical News January 12, 1990.

Adverse reactions to aspartame: Reply. Journal of Applied Nutrition 1989; 41: 40-41.

The licensing of Simplesse®: An open letter to the FDA. Journal of Applied Nutrition 1989; 41:42-43.

Aspartame (NutraSweet®): Is It Safe? Philadelphia, The Charles Press, 1989.

Is Aspartame (NutraSweet®) Safe? Public Health and Legal Challenges. In Proceedings for 30th Anniversary Conference on Legal Medicine. Orlando, Florida, March 15-17, 1990, 64-84.

The author's response to Robert H. Moser; The new issue of brain tumors. The Expert Witness Journal 1990:2 (No. 4, April): 15-17.

Clinicians and "clinical ecology": Commentary on a position paper. Journal of the Florida Medical Association 1990; 75:533-534.

Obstacles confronting consumer advocates: An overview of health-related issues. Trauma 1990; 31 (#5): 55-63.

Aspartame, tryptophan, and other amino acids as potentially hazardous experiments. (Letter) Southern Medical Journal 1990; 83:1110-1111.

Does aspartame (NutraSweet®) cause human brain cancer? Clinical Research 1990; 38:798A.

If it aches, avoid aspartame. (Letter) Cortlandt Forum 1990 October, 44.

Aspartame-associated confusion and memory loss. Townsend Letter for Doctors 1991; June, 442-443.

Aspartame (NutraSweet®): Is It Safe? The Author Responds. (Letter) Townsend Letter for Doctors 1991; October: 788-789.

Does aspartame cause human brain cancer? Journal of Advancement in Medicine 1991;4 (Winter):231-241.

Joint pain associated with aspartame use. Townsend Letter For Doctors 1991; May: 375-376.

Myasthenia gravis associated with aspartame use. Townsend Letter For Doctors 1991; August: 699-700.

Aspartame: Is it safe? Interview with H.J. Roberts, M.D. Mastering Food Allergies 1992;7 (#1), 3-6.

INDEX

B

C

D

E

Robinson, "Pat" (Reverend) 73, 74, 247

Russia 46, 172, 173

S

Saccharin 23, 58, 59, 108, 125, 212, 213

Satiety center 149

Schultz, Dorothy 183, 184

Sciencegate 76

Scribbling 243, 244

Serendipity 96, 97, 182-189

Silicone gel implants 161, 262

Skin 14, 33, 125, 130, 216

Skinner, Samuel 213

Sleepiness, Narcolepsy 14, 199

Slurred speech 14

Smith, Janet (Professor) 52-54, 158, 159

Social histories 190-194

Stereoisomers 167, 168, 170, 174, 189, 264, 272

Stevia rebaudiona 229, 230

Stoddard, Mary Nash 86, 87-89, 125, 126, 170, 222-231, 236, 250, 256, 276-282

Sugar (sucrose) 58, 206

   consumption 108

   industry 77, 78

   yearly consumption 108

Suicide 131

Sulfite sensitivity 207

Sweet terms 79-84

Sweeteners 82-84, 108, 208

Symptoms

   cost of studies 17,

   produced by rechallenge 13

Syndromes 205-209

Synthetic foods-additives 20, 83

T

Taylor, Larry 171, 230, 231

Television 72, 73, 74, 226-230, 249-251

Thalidomide 96

The Stuff 32

Think thin obsession (TTO) 144, 145

Thirst, increased 14, 247

Tinnitus 14

Toll 17

Tremors 14

"Truth-Seekers Seven" 179, 180

Tumors 20, 22, 95, 96, 210-217, 267, 268

U

U.S. Senate hearings 257, 258

University of North Texas seminar 68-70, 74, 97, 179, 180, 232, 238, 253

Urinary symptoms 14, 130, 131, 148, 285

W

Weight gain, paradoxical 14, 40, 92, 148, 149, 247, 255

Weight loss 14, 18, 141-161

   hazards of 156-159

Weight-loss books 153

Weight-loss entrepreneurs 152, 153

Welch, Raquel 98

Withdrawal symptoms 147

Women 118